Randomization, Masking, and Allocation Concealment

Chapman & Hall/CRC Biostatistics Series

Editor-in-Chief

Shein-Chung Chow, Ph.D., Professor, Department of Biostatistics and Bioinformatics, Duke University School of Medicine, Durham, North Carolina

Series Editors

Byron Jones, Biometrical Fellow, Statistical Methodology, Integrated Information Sciences, Novartis Pharma AG, Basel, Switzerland

Jen-pei Liu, Professor, Division of Biometry, Department of Agronomy, National Taiwan University, Taipei, Taiwan

Karl E. Peace, Georgia Cancer Coalition, Distinguished Cancer Scholar, Senior Research Scientist and Professor of Biostatistics, Jiann-Ping Hsu College of Public Health, Georgia Southern University, Statesboro, Georgia

Bruce W. Turnbull, Professor, School of Operations Research and Industrial Engineering, Cornell University, Ithaca, New York

Published Titles

Adaptive Design Methods in Clinical Trials, Second Edition
Shein-Chung Chow and Mark Chang

Adaptive Designs for Sequential Treatment Allocation
Alessandro Baldi Antognini
and Alessandra Giovagnoli

Adaptive Design Theory and Implementation Using SAS and R, Second Edition
Mark Chang

Advanced Bayesian Methods for Medical Test Accuracy
Lyle D. Broemeling

Analyzing Longitudinal Clinical Trial Data: A Practical Guide
Craig Mallinckrodt and Ilya Lipkovich

Applied Biclustering Methods for Big and High-Dimensional Data Using R
Adetayo Kasim, Ziv Shkedy,
Sebastian Kaiser, Sepp Hochreiter,
and Willem Talloen

Applied Meta-Analysis with R
Ding-Geng (Din) Chen and Karl E. Peace

Applied Surrogate Endpoint Evaluation Methods with SAS and R
Ariel Alonso, Theophile Bigirumurame,
Tomasz Burzykowski, Marc Buyse,
Geert Molenberghs, Leacky Muchene,
Nolen Joy Perualila, Ziv Shkedy,
and Wim Van der Elst

Basic Statistics and Pharmaceutical Statistical Applications, Second Edition
James E. De Muth

Bayesian Adaptive Methods for Clinical Trials
Scott M. Berry, Bradley P. Carlin,
J. Jack Lee, and Peter Muller

Bayesian Analysis Made Simple: An Excel GUI for WinBUGS
Phil Woodward

Bayesian Designs for Phase I–II Clinical Trials
Ying Yuan, Hoang Q. Nguyen,
and Peter F. Thall

Bayesian Methods for Measures of Agreement
Lyle D. Broemeling

Bayesian Methods for Repeated Measures
Lyle D. Broemeling

Bayesian Methods in Epidemiology
Lyle D. Broemeling

Bayesian Methods in Health Economics
Gianluca Baio

Bayesian Missing Data Problems: EM, Data Augmentation and Noniterative Computation
Ming T. Tan, Guo-Liang Tian,
and Kai Wang Ng

Published Titles

Bayesian Modeling in Bioinformatics
Dipak K. Dey, Samiran Ghosh,
and Bani K. Mallick

Benefit-Risk Assessment in Pharmaceutical Research and Development
Andreas Sashegyi, James Felli,
and Rebecca Noel

Benefit-Risk Assessment Methods in Medical Product Development: Bridging Qualitative and Quantitative Assessments
Qi Jiang and Weili He

Bioequivalence and Statistics in Clinical Pharmacology, Second Edition
Scott Patterson and Byron Jones

Biosimilar Clinical Development: Scientific Considerations and New Methodologies
Kerry B. Barker, Sandeep M. Menon,
Ralph B. D'Agostino, Sr., Siyan Xu, and Bo Jin

Biosimilars: Design and Analysis of Follow-on Biologics
Shein-Chung Chow

Biostatistics: A Computing Approach
Stewart J. Anderson

Cancer Clinical Trials: Current and Controversial Issues in Design and Analysis
Stephen L. George, Xiaofei Wang,
and Herbert Pang

Causal Analysis in Biomedicine and Epidemiology: Based on Minimal Sufficient Causation
Mikel Aickin

Clinical and Statistical Considerations in Personalized Medicine
Claudio Carini, Sandeep Menon, and Mark Chang

Clinical Trial Data Analysis Using R
Ding-Geng (Din) Chen and Karl E. Peace

Clinical Trial Data Analysis Using R and SAS, Second Edition
Ding-Geng (Din) Chen, Karl E. Peace,
and Pinggao Zhang

Clinical Trial Methodology
Karl E. Peace and Ding-Geng (Din) Chen

Clinical Trial Optimization Using R
Alex Dmitrienko and Erik Pulkstenis

Cluster Randomised Trials: Second Edition
Richard J. Hayes and Lawrence H. Moulton

Computational Methods in Biomedical Research
Ravindra Khattree and Dayanand N. Naik

Computational Pharmacokinetics
Anders Källén

Confidence Intervals for Proportions and Related Measures of Effect Size
Robert G. Newcombe

Controversial Statistical Issues in Clinical Trials
Shein-Chung Chow

Data Analysis with Competing Risks and Intermediate States
Ronald B. Geskus

Data and Safety Monitoring Committees in Clinical Trials, Second Edition
Jay Herson

Design and Analysis of Animal Studies in Pharmaceutical Development
Shein-Chung Chow and Jen-pei Liu

Design and Analysis of Bioavailability and Bioequivalence Studies, Third Edition
Shein-Chung Chow and Jen-pei Liu

Design and Analysis of Bridging Studies
Jen-pei Liu, Shein-Chung Chow,
and Chin-Fu Hsiao

Design & Analysis of Clinical Trials for Economic Evaluation & Reimbursement: An Applied Approach Using SAS & STATA
Iftekhar Khan

Design and Analysis of Clinical Trials for Predictive Medicine
Shigeyuki Matsui, Marc Buyse,
and Richard Simon

Design and Analysis of Clinical Trials with Time-to-Event Endpoints
Karl E. Peace

Design and Analysis of Non-Inferiority Trials
Mark D. Rothmann, Brian L. Wiens,
and Ivan S. F. Chan

Published Titles

Difference Equations with Public Health Applications
Lemuel A. Moyé and Asha Seth Kapadia

DNA Methylation Microarrays: Experimental Design and Statistical Analysis
Sun-Chong Wang and Arturas Petronis

DNA Microarrays and Related Genomics Techniques: Design, Analysis, and Interpretation of Experiments
David B. Allison, Grier P. Page,
T. Mark Beasley, and Jode W. Edwards

Dose Finding by the Continual Reassessment Method
Ying Kuen Cheung

Dynamical Biostatistical Models
Daniel Commenges and
Hélène Jacqmin-Gadda

Elementary Bayesian Biostatistics
Lemuel A. Moyé

Emerging Non-Clinical Biostatistics in Biopharmaceutical Development and Manufacturing
Harry Yang

Empirical Likelihood Method in Survival Analysis
Mai Zhou

Essentials of a Successful Biostatistical Collaboration
Arul Earnest

Exposure–Response Modeling: Methods and Practical Implementation
Jixian Wang

Frailty Models in Survival Analysis
Andreas Wienke

Fundamental Concepts for New Clinical Trialists
Scott Evans and Naitee Ting

Generalized Linear Models: A Bayesian Perspective
Dipak K. Dey, Sujit K. Ghosh, and
Bani K. Mallick

Handbook of Regression and Modeling: Applications for the Clinical and Pharmaceutical Industries
Daryl S. Paulson

Inference Principles for Biostatisticians
Ian C. Marschner

Interval-Censored Time-to-Event Data: Methods and Applications
Ding-Geng (Din) Chen, Jianguo Sun,
and Karl E. Peace

Introductory Adaptive Trial Designs: A Practical Guide with R
Mark Chang

Joint Models for Longitudinal and Time-to-Event Data: With Applications in R
Dimitris Rizopoulos

Measures of Interobserver Agreement and Reliability, Second Edition
Mohamed M. Shoukri

Medical Biostatistics, Fourth Edition
A. Indrayan

Meta-Analysis in Medicine and Health Policy
Dalene Stangl and Donald A. Berry

Methods in Comparative Effectiveness Research
Constantine Gatsonis and Sally C. Morton

Mixed Effects Models for the Population Approach: Models, Tasks, Methods and Tools
Marc Lavielle

Modeling to Inform Infectious Disease Control
Niels G. Becker

Modern Adaptive Randomized Clinical Trials: Statistical and Practical Aspects
Oleksandr Sverdlov

Monte Carlo Simulation for the Pharmaceutical Industry: Concepts, Algorithms, and Case Studies
Mark Chang

Multiregional Clinical Trials for Simultaneous Global New Drug Development
Joshua Chen and Hui Quan

Multiple Testing Problems in Pharmaceutical Statistics
Alex Dmitrienko, Ajit C. Tamhane,
and Frank Bretz

Published Titles

Noninferiority Testing in Clinical Trials: Issues and Challenges
Tie-Hua Ng

Optimal Design for Nonlinear Response Models
Valerii V. Fedorov and Sergei L. Leonov

Patient-Reported Outcomes: Measurement, Implementation and Interpretation
Joseph C. Cappelleri, Kelly H. Zou, Andrew G. Bushmakin, Jose Ma. J. Alvir, Demissie Alemayehu, and Tara Symonds

Quantitative Evaluation of Safety in Drug Development: Design, Analysis and Reporting
Qi Jiang and H. Amy Xia

Quantitative Methods for HIV/AIDS Research
Cliburn Chan, Michael G. Hudgens, and Shein-Chung Chow

Quantitative Methods for Traditional Chinese Medicine Development
Shein-Chung Chow

Randomization, Masking, and Allocation Concealment
Vance W. Berger

Randomized Clinical Trials of Nonpharmacological Treatments
Isabelle Boutron, Philippe Ravaud, and David Moher

Randomized Phase II Cancer Clinical Trials
Sin-Ho Jung

Repeated Measures Design with Generalized Linear Mixed Models for Randomized Controlled Trials
Toshiro Tango

Sample Size Calculations for Clustered and Longitudinal Outcomes in Clinical Research
Chul Ahn, Moonseong Heo, and Song Zhang

Sample Size Calculations in Clinical Research, Third Edition
Shein-Chung Chow, Jun Shao, Hansheng Wang, and Yuliya Lokhnygina

Statistical Analysis of Human Growth and Development
Yin Bun Cheung

Statistical Design and Analysis of Clinical Trials: Principles and Methods
Weichung Joe Shih and Joseph Aisner

Statistical Design and Analysis of Stability Studies
Shein-Chung Chow

Statistical Evaluation of Diagnostic Performance: Topics in ROC Analysis
Kelly H. Zou, Aiyi Liu, Andriy Bandos, Lucila Ohno-Machado, and Howard Rockette

Statistical Methods for Clinical Trials
Mark X. Norleans

Statistical Methods for Drug Safety
Robert D. Gibbons and Anup K. Amatya

Statistical Methods for Healthcare Performance Monitoring
Alex Bottle and Paul Aylin

Statistical Methods for Immunogenicity Assessment
Harry Yang, Jianchun Zhang, Binbing Yu, and Wei Zhao

Statistical Methods in Drug Combination Studies
Wei Zhao and Harry Yang

Statistical Testing Strategies in the Health Sciences
Albert Vexler, Alan D. Hutson, and Xiwei Chen

Statistics in Drug Research: Methodologies and Recent Developments
Shein-Chung Chow and Jun Shao

Statistics in the Pharmaceutical Industry, Third Edition
Ralph Buncher and Jia-Yeong Tsay

Survival Analysis in Medicine and Genetics
Jialiang Li and Shuangge Ma

Theory of Drug Development
Eric B. Holmgren

Translational Medicine: Strategies and Statistical Methods
Dennis Cosmatos and Shein-Chung Chow

Randomization, Masking, and Allocation Concealment

Edited by
Vance W. Berger

CRC Press
Taylor & Francis Group
Boca Raton London New York

CRC Press is an imprint of the
Taylor & Francis Group, an **informa** business

A CHAPMAN & HALL BOOK

CRC Press
Taylor & Francis Group
6000 Broken Sound Parkway NW, Suite 300
Boca Raton, FL 33487-2742

First issued in paperback 2020

© 2018 by Taylor & Francis Group, LLC
CRC Press is an imprint of Taylor & Francis Group, an Informa business

No claim to original U.S. Government works

ISBN-13: 978-0-367-73561-6 (pbk)
ISBN-13: 978-1-138-03364-1 (hbk)

Library of Congress Cataloging-in-Publication Data

Names: Berger, Vance, editor.
Title: Randomization, masking, and allocation concealment / [edited by] Vance Berger.
Description: Boca Raton : Taylor & Francis, a CRC title, part of the Taylor & Francis imprint, a member of the Taylor & Francis Group, the academic division of T&F Informa plc, 2018. | Includes bibliographical references.
Identifiers: LCCN 2017030500 | ISBN 9781138033641 (hardback : acid-free paper)
Subjects: LCSH: Clinical trials--Statistical methods. | Clinical trials--Moral and ethical aspects.
Classification: LCC R853.C55 R35 2018 | DDC 610.72/4--dc23
LC record available at https://lccn.loc.gov/201703050

Visit the Taylor & Francis Web site at
http://www.taylorandfrancis.com

and the CRC Press Web site at
http://www.crcpress.com

Contents

Editor...xi

Contributors..xiii

1. Randomization and Bias in Historical Perspective.............................1
   J. Rosser Matthews

2. Proper Randomization Reduces the Chance of Wasted
   Biomedical Research..9
   Arturo Martí-Carvajal

3. Sympathetic Bias: A Neglected Source of Selection Bias....................29
   William C. Grant

4. The Alleged Benefits of Unrestricted Randomization 39
   Vance W. Berger

5. Restricted Randomization: Pros and Cautions 51
   Jonathan Chipman

6. Evolution of Restricted Randomization with Maximum
   Tolerated Imbalance...61
   Wenle Zhao

7. Evaluating the Evaluation...83
   Adriana C. Burgos and Ross J. Kusmick

8. Selection Bias in Studies with Unequal Allocation 89
   Olga M. Kuznetsova

9. Unrecognized Dual Threats to Internal Validity Relating
   to Randomization 123
   Vance W. Berger, Adriana C. Burgos, and Omolola A. Odejimi

10. Testing for Second-Order Selection Bias Effect in Randomised
    Controlled Trials Using Reverse Propensity Score (RPS).................141
    Steffen Mickenautsch and Bo Fu

11. **The Berger–Exner Test to Detect Third-Order Selection Bias
    in the Presence of a True Treatment Effect** 159
    Steffen Mickenautsch, Bo Fu, and Vance W. Berger

12. **Adjusting for and Detection of Selection Bias in Randomized
    Controlled Clinical Trials** 171
    Lieven N. Kennes

13. **Randomization and the Randomization Test: Two Sides
    of the Same Coin** 185
    Patrick Onghena

14. **Randomization Tests or Permutation Tests? A Historical
    and Terminological Clarification** 209
    Patrick Onghena

15. **Flexible Minimization: Synergistic Solution for Selection Bias** 229
    Donald R. Taves

Index 241

Editor

Vance W. Berger, PhD, earned a doctoral degree in Statistics from Rutgers University. His professional career has included work in the pharmaceutical industry (Janssen Research Foundation, Theradex, and some consulting for Pfizer), work in two centers of the Food and Drug Administration (Drugs and Biologics), and review work for a number of statistical and medical journals. An active researcher, Dr. Berger wrote a book on the design and analysis of randomized clinical trials (focusing on randomization methods). Dr. Berger has also authored numerous book chapters and scientific articles appearing in the peer-reviewed literature and has presented numerous invited lectures on this topic. Dr. Berger was the recipient of the 2006 Gertrude Cox Award, recognizing "A statistician making significant contributions to statistical practice" by the Washington Statistical Society.

Contributors

Adriana C. Burgos
Universidad Privada del Valle
Cochabamba, Bolivia

Jonathan Chipman
Department of Biostatistics
Vanderbilt School of Medicine
Nashville, Tennessee

Bo Fu
Data and Statistical Science
AbbVie Inc
Chicago, Illinois

William C. Grant
Department of Economics
James Madison University
Harrisonburg, Virginia

Lieven N. Kennes
Department of Economics
University of Applied Sciences
   Stralsund
Stralsund, Germany

Ross J. Kusmick
Rutgers University
New Brunswick, New Jersey

Olga M. Kuznetsova
Late Development Statistics
Merck & Co., Inc.
Rahway, New Jersey

Arturo Martí-Carvajal
Universidad de Carabobo
Valencia, Venezuela

Steffen Mickenautsch
SYSTEM Initiative/Department
   of Community Dentistry
Faculty of Health Sciences,
   University of the Witwatersrand
Johannesburg, South Africa

Omolola A. Odejimi
Bachelor of Science
Department of Psychology
Portland State University
Portland, Oregon

Patrick Onghena
Faculty of Psychology and
   Educational Sciences
KU Leuven
Leuven, Belgium

J. Rosser Matthews
Professional Writing Program
English Department
University of Maryland
College Park, Maryland

Donald R. Taves
Department of Oral Biology
University of Washington
Seattle, Washington

Wenle Zhao
Medical University of South
   Carolina
Charleston, South Carolina

1

Randomization and Bias in Historical Perspective

J. Rosser Matthews

CONTENTS

1.1 Early Twentieth-Century Discussions of Bias and "Randomization"
    in Clinical Research...2
1.2 Austin Bradford Hill and the Streptomycin Trial in Treating
    Tuberculosis (1948)..3
1.3 "Lessons Learned" and Permuted Block Design 5
1.4 Randomization and the Ethics of Clinical Trials 6
References..7

Randomized controlled trials (RCTs) are justified by many considerations—both methodological and ethical. From a methodological standpoint, the assignment of subjects to treatment arms based on random criteria is designed to prevent systematic differences from influencing the outcomes; potential differences between the arms of the trial will be "washed out" via the random procedure. Similarly, the double-blind element is designed to prevent the subjective biases of the researcher from influencing the outcome.[1] From an ethical standpoint, the use of the random element is tied up with broader ideas of fairness; just as in sporting events and political debates, a chance element (e.g., a flip of a coin) is used to ensure that no side gets an unfair advantage. Also, since the results from a randomized study are more credible than observational studies without randomization, the determination of safety and efficacy might be accomplished more quickly than through unsystematic observation; when interventions pose potentially significant health risks, this is clearly a socially beneficial outcome.

But problems still exist when researchers seek to operationalize these procedures in practice. In particular, if researchers can deduce the randomization procedure, then there is the possibility of still assigning subjects in ways that will bias the final outcome. Over the years, strategies for addressing this problem have evolved; however, the underlying rationale of preventing bias remains. In this chapter, I will examine the history of how bias and randomization have been dealt with to assess the "lessons learned" from past experience.

1.1 Early Twentieth-Century Discussions of Bias and "Randomization" in Clinical Research

As early as the first decade of the twentieth century, there was overt discussion of bias and how it could be prevented—as illustrated by published discussions about anti-typhoid inoculation by the bacteriologist Sir Almroth Wright and the statistician Karl Pearson. In his 1904 book *A Short Treatise on Anti-Typhoid Inoculation*, Wright outlined how he believed the statistical arguments on this issue should be assessed. He claimed that there should be a control group "which ought to correspond with the inoculated group in all points save only in the circumstance of inoculation." Also, he claimed that small errors in the data could be overlooked because a large number of observations would mean "all chance errors... are spontaneously eliminated." However, he warned that, if an error always favored one conclusion over another, it would not necessarily be eliminated by the accumulation of a large number of observations.[2] Although he did not use the term "bias," he clearly acknowledged the possibility of systematic distortion when reporting statistical findings. However, he still claimed that his medical judgment could trump statistical rules: "The plain everyday man will find it possible to reconcile the demands of his statistical conscience with the demands of his practical life. He will neglect the mint and anise and cummin of statistical criticism while holding fast to weightier principles of the statistical law."[2]

Wright's methodological views were questioned by Karl Pearson. In a letter to an official in the British War Office, Pearson noted how the *voluntary* nature of inoculation could lead to distorted results. He commented that "the regiment being divided for special duties, companies might very easily run different risks... May not... the more careful men have been inoculated, just as the more cautious are vaccinated? Thus the average correlation between inoculation and immunity might only mean a correlation between greater caution and immunity." To prevent this, Pearson advocated an experimental protocol in which every alternative man in a regiment of 800 be inoculated so that "we can exhibit your results in correlative form, showing a distinct relation between inoculation and immunity."[3] Even though the War Office recommended against such as alternation procedure (since inoculation was voluntary),[4] Pearson did publish an article in the *British Medical Journal* in which he computed the correlation coefficient for this procedure so that it could be compared to other widely used procedures such as smallpox vaccination.[5] What this exchange shows is that both Wright and Pearson acknowledged that chance and bias could play a role when data are collected; however, neither advocated for a method of completely "random" assignment as a way to address the bias issue.

In the following decades, Pearson's recommendation—alternate allocation of patients to the experimental and control groups—would become a prominent feature of clinical trial design. As Iain Chalmers has noted, the approach was used in studies on plague and cholera in India in the first decade of the century. In the second and third decades of the century, alternate allocation was used in controlled trials of serum treatment for pneumonia in the United States.[6]

In later studies, the word "random" was explicitly used to discuss subject assignment. However, it usually still referred to alternative allocation. Illustrative of this semantic ambiguity is the 1938 study reported by Diehl et al., which involved testing the efficacy of a vaccine for the common cold among college students at the University of Minnesota. The assignment procedure was described in the following manner:

> At the beginning of each year of the study students were assigned at random and without selection to a control or to an experimental group. The students in the control groups were treated in exactly the same manner as those in the experimental groups but received placebos instead of vaccine. All students thought that they were receiving vaccine and so had an unprejudiced attitude toward the study. Even the physicians who saw the students at the health service when they contracted colds during the period of the study had no information as to which group they represented.[7]

While this description has a very "modern" sound to it (aside from the unethical practice of leaving all students under the impression that they were getting the vaccine), Armitage has documented that Diehl actually used alternate allocation when assigning the subjects. He had used alternation in earlier common cold trials, and in a speech from 1941, Diehl declared that "at the beginning of the [1938] study, students who volunteered to take these treatments were assigned alternatively and without selection to control groups and experimental groups."[8]

1.2 Austin Bradford Hill and the Streptomycin Trial in Treating Tuberculosis (1948)

By general consensus, the British statistician Austin Bradford Hill (1897–1991) is regarded as the "father" of the truly *randomized* controlled trials. Before World War II, Hill supported alternation as a method of patient assignment. In the first edition of his 1937 textbook *Principles of Medical Statistics*, Hill wrote that alternative allocation was "often satisfactory" because "we can fairly rely upon this random allotment of patient to equalize in the two groups the distribution of other characteristics."[6] Reflecting on this recommendation over half a century later, Bradford Hill noted that he "was trying to persuade the doctors to come into controlled trials in the very simplest form and I might have scared them off... I thought it would be better to get doctors to walk first, before I tried to get them to run."[9]

In World War II, Hill would be provided with an opportunity to develop a new patient assignment mechanism when he was appointed to a committee that was going to administer a clinical trial of the drug streptomycin to treat tuberculosis. Streptomycin had been developed by researchers in the United States, and the preliminary evidence appeared to suggest that patients might improve when given the drug. To determine whether this improvement was related to receiving streptomycin, a clinical trial was launched in the United Kingdom by the Medical Research Council. Because of wartime shortages, many tuberculosis patients would still have had to go untreated—whether a controlled clinical trial was executed or not. This empirical fact allayed the ethical qualms of those conducting the trial on the issue of having a placebo control group. As D.D. Reid observed in 1950, "Our genteel poverty thus paid a scientific dividend by quieting any doubts we might have about the ethics of controlled trials of streptomycin in pulmonary tuberculosis."[10]

Instead of alternate allocation, Hill used a randomization procedure. Each participating hospital was given a series of numbered envelopes (with a separate list for each gender), and each envelope contained a card that was marked either "S" (streptomycin) or "C" (control); the order of "S" and "C" was determined by a series of random sampling numbers. When each new participant was enrolled, the next numbered envelope would be opened, and the patient would be assigned to the designated group.[11] Later, Hill would emphasize that the rationale for this methodological switch was to eliminate the potential for bias:

> The aim, therefore, is to allocate the patients to the "treatment" and "control" groups in such a way that the two groups are initially equivalent in all respects relevant to the inquiry... In many trials this allocation has been successfully made by putting patients, as they present themselves, alternately into the treatment and control groups. Such a method may, however, be insufficiently random... if the clinician involved knows whether the patient, if accepted, will pass to the treatment or the control group. By such knowledge he may be biased... For this reason, recent British trials have avoided the alternating method and adopted the use of random sampling numbers; in addition, the allocation of the patient to treatment or control is kept secret from the clinician until after he has made his decision upon the patient's admission.[12]

The trial consisted of 107 patients, all of whom suffered from bilateral pulmonary tuberculosis. Fifty-five of the subjects were assigned to the group to receive doses of streptomycin on a regular basis (plus the standard therapy of bed rest) and the other 52 subjects (the control group) received only bed rest. At the end of the trial, four of the patients in the streptomycin group had died (7%), whereas 14 in the control group had died (27%); the difference in results was statistically significant.[13]

1.3 "Lessons Learned" and Permuted Block Design

What this brief history illustrates is that the process of clinical trial design consists in continual methodological refinement to achieve results less influenced by bias. Initially, the principle of alternate allocation was used on the theory that, if a sufficiently large number of observations were recorded, any errors might be "cancelled out." However, as even Almroth Wright observed, this strategy would not preclude bias if the errors always favored one outcome over another. This would appear to suggest the need for preventing trial administrators from being able to determine which patient would be assigned to each arm of the study. One strategy to achieve this was to replace alternate assignment with some form of a randomization procedure—such as relying on a system of random numbers like Austin Bradford Hill.

But randomization is not synonymous with arbitrary; there still should be a "structure" to ensure that the intervention and the control groups are comparable—for example, that they have the same number of members. Equal group size is not assured with a completely random sequence (a fair coin can come up heads successively after several tosses by chance). Historically, one way to deal with this issue has been to employ permuted block design. According to this strategy, patients are assigned to the treatment or control arm in blocks (e.g., of four patients, six patients, etc.), and each block will be a different possible permutation of the ways that the two or more interventions could be sequenced. Within each block, however, an equal number of patients will be assigned to each trial arm. By randomly selecting a different permutation of the block design when one block has been completed, researchers are attempting to achieve both variability (from block to block) and order (equal assignment).

While permuted block design might appear to be "the best of all possible worlds," it is actually opening the door for the same type of bias as alternative assignment—albeit at a higher level of conceptual sophistication. If researchers know the size of each permutation block, they will not know how the first patients will be assigned; however, they can deduce how the last patients in a block will be assigned—by process of elimination. Consequently, there are still ways for the knowledge of the clinicians to bias the outcome. Since they might be able to deduce how a *specific subset* will be assigned based on their location in the permutation block, this could be enough to undermine the results—if those specific patients were assigned in systematically distorting ways. What this historical analogy suggests is that just as alternative allocation had to be replaced with an early form of randomization (in the service of eliminating bias), so likewise must the permuted block strategy be reassessed to ensure continued methodological rigor in the execution of randomized controlled clinical trials in the future.

1.4 Randomization and the Ethics of Clinical Trials

Using randomization as a way to minimize bias is often discussed in the context of ensuring methodological rigor. As the above discussion indicates, this is undeniably important. However, it is also important to view these practices through an ethical lens. As a form of experimentation on a human population, there will always be a risk (and potential benefit) from electing to participate in a clinical trial. Consequently, designers of clinical trials have a duty not only to ensure that a trial is executed in a scientific manner, but also to ensure that risks and benefits are spread equitably and administered impartially.

Ironically, impartiality is often achieved through a randomization procedure—such as flipping a coin. In competitive sports, a coin flip is often used to decide which team has the ball initially, or is the first at bat. In political debates, a coin flip often determines which candidate gets to make the opening and/or closing statement, or is asked the first question. Implicit in these practices is the view that, when dealing with two or more *credible and competing* alternatives, the fairest procedure to determine which side should be "up" first should depend on a random outcome—such as the result of a single flip of a fair coin.

Viewed from this perspective, the use of randomization to assign participants in a clinical trial to different arms of the trial seems self-evident. Just as in sports or politics, there are only a finite number of possible outcomes in a clinical trial (either there is a significant difference in the outcomes of the various treatment arms or there isn't). Similarly, in sports, one of the two competing teams will win and, in politics, one of the two (or more) candidates will win. Because of the structural similarity in the organization of sports, politics, and clinical trials (they are all ruled-governed activities with explicitly stated criteria that determine which outcome is accepted as dispositive), the use of randomization techniques would seem to be justified in all three instances based on the same rationale—the association of chance assignment with procedural fairness. Ultimately, what this means is that continuing to strive to eliminate bias in trial design is an end that has both a scientific and an ethical justification.

References

1. Sibbald B, Roland M (1998). Understanding controlled trials: Why are randomised controlled trials important? *BMJ* 316:201.
2. Matthews JR (1995). *Quantification and the Quest for Medical Certainty*: pp. 98–99. Princeton: Princeton University Press.
3. K. Pearson to R. J. S. Simpson, Karl Pearson Papers, May 24, 1904, Box 159/1. University College London.
4. Simpson to Pearson, Karl Pearson Papers, May 30, 1904, Box 159/1. University College London.
5. Pearson K (1904). Report on certain enteric fever inoculation statistics. *BMJ* 3:1243–1246.
6. Chalmers I (2013). UK Medical Research Council and multicenter clinical trials: From a damning report to international recognition. JLL Bulletin: Commentaries on the history of treatment evaluation (http://www.jameslind library.org/articles/uk-medical-research-council-and-multicentre-clinical-trials -from-a-damning-report-to-international-recognition/).
7. Diehl HS, Baker AB, Cowan DW (1938). Cold vaccines: An evaluation based on a controlled study. *JAMA* 1168–1173.
8. Armitage P (2002). Randomisation and alternation: A note on Diehl et al. JLL Bulletin: Commentaries on the history of treatment evaluation (http:// www.jameslindlibrary.org/articles/randomisation-and-alternation-a-note-on -diehl-et-al/).
9. Hill AB (1990). Memories of the British streptomycin trial in tuberculosis: The first randomized clinical trial. *Controlled Clinical Trials* 11:77–79. Quoted in Chalmers (2013).
10. Reid DD (1950). Statistics in clinical research. *Annals of the New York Academy of Sciences* 52:931–934.
11. Crofton J (2006). The MRC randomized trial of streptomycin and its legacy: A view from the clinical front line. *J R Soc Med* 99:531–534.
12. Hill AB (1951). The clinical trial. British Medical Bulletin 7:278–282.
13. Medical Research Council (1948). Streptomycin treatment of pulmonary tuberculosis: A Medical Research Council investigation. *BMJ* 2:769–782.

2

Proper Randomization Reduces the Chance of Wasted Biomedical Research

Arturo Martí-Carvajal

CONTENTS

2.1 Introduction...9
2.2 What Does Randomization Mean?..10
2.3 What Is Randomization?...11
2.4 Why Is Randomization Crucial?..12
2.5 What Type of Randomization Process Are You Looking for?..............13
2.6 Is There an Association between Confusion Bias
     and Inappropriate Randomization Process?...14
2.7 How Should the Randomization Section Be Reported?........................21
2.8 Colophon...23
References..24

2.1 Introduction

In 1925, Ronald A. Fisher, in his book "Statistical Methods for Research Workers," said "...to discriminate between those conclusions which flow from the nature of the observations themselves, and those which are due solely to faulty methods of estimation" (Fisher 1925), it could be understood as a prelude to what is known almost a century later as waste research (Ioannidis 2014). In another book, *The Design of Experiments*, Dr. Fisher gifts us this sentence "Randomisation properly carried out, ... ensures that the estimates of error will take proper care of all such causes of different growth rates, and relieves the experimenter from the anxiety of considering and estimating the magnitude of the innumerable causes by which his data may be disturbed" (Fisher 1935). These two reflections by R.A. Fisher have not been used properly by several researchers, but rather have been used as a sort of franchise or trademark for falsely improving the quality of their clinical trials. It has been pointed out that more than half are not randomized at all (Ioannidis 2014), although, in 1948, the first randomized clinical trial (RCT) was conducted, which compared streptomycin versus placebo in people with pulmonary tuberculosis (Medical Research Council 1948). Despite this, there seems to be a deliberate ignorance for conducting trials without a proper randomization process; that is, only use the word "randomized" as a label, tag, or mark.

In the age of evidence-based medicine (Sackett et al. 1996), which has been considered as a dialogue between Socrates and Hippocrates (Howick 2011), it is fundamental to understand the philosophy of RCTs: the best evidence* to evaluate the clinical effectiveness of an intervention. The *Stanford Encyclopedia of Philosophy* cites to evidence as "The concept of evidence is central to both epistemology and the philosophy of science"... and "Evidence, whatever else it is, is the kind of thing which can make a difference to what one is *justified* in believing or (what is often, but not always, taken to be the same thing) what it is *reasonable* for one to believe" (Kelly 2016).

As an author of systematic reviews of healthcare interventions at the Cochrane Collaboration (Cassels 2015), I am in keeping touch with the "movie stars" in any systematic review: the clinical trials. The first question will be: "Mr. Clinical Trial, is randomized your last name?" If "Mr. Clinical Trial" answers "Yes!", we must be sure that it is true. How to do that? We must critically appraise that "Yes" due to it is very often the kind of "yes" that is a false positive! Mr. Clinical Trial might use a fake identity card! Therefore, he is a dummy randomized. Based on the above, many questions emerge. *First*, why is the RCTs the best evidence to assess the clinical effectiveness of an intervention? *Second*, why is the first question whether the clinical trial was randomized? *Third*, is randomization what confers glamor on the clinical trial? Why is it so crucial? What does randomization mean?

Let us start with the last question, What does randomization mean?

2.2 What Does Randomization Mean?

According to the *Dictionary of Epidemiology* (Porta 2014), *randomization* is "allocation[†] of individuals to groups in a clinical trial (e.g., intervention

---

* *Online Etimology Dictionary*, Evidence, "appearance from which inferences may be drawn," from Old French *evidence*, from Late Latin *evidentia* "proof," in classical Latin "distinction, vivid presentation, clearness" in rhetoric, from stem of Latin *evidens* "obvious, apparent," late 14c., from Old French *evident* and directly from Latin *evidentem* (nominative *evidens*) "perceptible, clear, obvious, apparent" from *ex* "out, out of, fully" (see *ex-*) + *videntem* (nominative *videns*), present participle of *videre* "to see." https://goo.gl/uPbalb Accessed 12 January 2017).
† Allocation. Mid 15c., "authorization," from Middle French *allocacion*, from Medieval Latin *allocationem* (nominative allocatio), noun of action from past participle stem of allocate, "allot," from assimilated form of Latin *ad* "to" + *locare* "to place," from locus "a place." http://www.etymonline.com.

and control) by chance.*" As randomization is applied to a clinical trial (i.e., "randomized" clinical trial), it raises another question, what is chance? The meaning of it from the same dictionary is "frequency probability." The *Stanford Encyclopedia of Philosophy* states that "Chance and randomness are closely related. So much so, in fact, that to say an event happened by chance is near enough synonymous in ordinary English with saying it happened randomly" (Eagle 2016). Therefore, it would be mandatory to ask what is random?[†] The *Dictionary of Epidemiology* (Porta 2014) defines random as "governed by chance; not completely determined by measurable factors." Summarizing, the human being's hands should not be immersed into allocation of participants to the comparison groups.

2.3 What Is Randomization?

Once we have known what *randomization* means, we must ask ourselves: How does the researcher randomize or should the assignment of participants to the comparison groups be truly by chance?

Let us start by saying that randomization is a unique process composed for two phases: generation of the allocation sequence and the randomness or concealment of the allocation sequence of that first phase (Cipriani et al. 2008; Viera and Bangdiwala 2007). As much as possible, these phases should be carried out in one step, simultaneously, because it reduces the likelihood of violating the generation of the allocation sequence. The crux matter of the randomization process is for the researcher to conduct both phases, but without knowing what the result of allocation was. In other words, randomness should be unpredictability (Eagle 2005). Hence, randomization should not be considered as a technique (Berger and Bears 2003).

There are numerous techniques to ensure that it is chance, not the will or desire of the researcher, that assigns participants to comparison groups (Berger 2006; Egbewale 2014; Higgins et al. 2011; Kim and Shin 2014; Schulz and Grimes 2002; White and Freedman 1978). Describing these techiniques is beyond the scope of this chapter. There are many websites (or software downloads) where one can get randomization for free (Adaptive Randomization 2017;

---

* According to *Online Etimology Dictionary*, Chance c. 1300, "something that takes place, what happens, an occurrence" (good or bad, but more often bad), from Old French *cheance* "accident, chance, fortune, luck, situation, the falling of dice" (12c., Modern French *chance*), from Vulgar Latin **cadentia* "that which falls out," a term used in dice, from neuter plural of Latin *cadens*, present participle of *cadere* "to fall," from PIE root **kad-* "to lay out, fall or make fall" https://goo.gl/fveWuQ (Accessed 12 January 2017).
† According to *Online Etimology Dictionary*, Random (adj), "having no definite aim or purpose," 1650s, from *at random* (1560s), "at great speed" (thus, "carelessly, haphazardly"). https://goo.gl/h5Cphq https://goo.gl/fveWuQ (Accessed 12 January 2017).

Random Allocation Software 2017; Random.org 2017; Randomization.com 2017; Research Randomizer 2017; Sealed Envelope 2017). These tools help carry out the randomization process in research sites with middle or low economic income, which impedes high-quality biomedical research.

2.4 Why Is Randomization Crucial?

The randomization process is important because it is based on clinical equipoise's principle* (Freedman 1987). According to this principle, "Clinical equipoise obtains when, for the population of patients from which subjects will be selected, the available clinical evidence does not favor one of the treatments being used over the others. In addition, it must be the case that there are no treatments available outside the trial that are better than those used in the trial. Satisfaction of these conditions seems to imply that the interests of research subjects will not be undermined in the service of collecting scientific information. If the available data do not favor any of the treatments being used, randomizing subjects seems as good a process as any other for choosing which treatment they receive" (Wendler 2016). As the intervention groups have *equi poise*, a judge will be required to elucidate what the best treatment is: randomization will be the judge. The opposite would be against ethics.

It is assumed that when a clinical trial is proposed to evaluate the clinical benefits of an intervention with respect to control, the researcher expects that the difference (if any) between the groups is attributed to the experimental intervention or study drug: Fisher's advice! As happens in a competition, the difference will be explained by the ability of competitors: doping is prohibited! This assumption implies a principle called *ceteris paribus*[†] (Lilienfeld 1982). Fulfilling *ceteris paribus* means reducing the likelihood of the confusion[‡] bias[§] owing to an inappropriate randomization process. It is

---

* Equipoise, "an equal distribution of weight," 1650s, a contraction of the phrase *equal poise* (1550s). http://www.etymonline.com.
† The Latin phrase "ceteris paribus" or "caeteris paribus"—literally meaning "other things being equal"—was used in a non-technical sense by Cicero. (Reutlinger, Alexander, Schurz, Gerhard and Hüttemann, Andreas, "Ceteris Paribus Laws," *The Stanford Encyclopedia of Philosophy* (Fall 2015 Edition), Edward N. Zalta (ed.), URL = https://plato.stanford.edu /archives/fall2015/entries/ceteris-paribus/. (Accessed 25 January 2017.)
‡ Confound C.1300, "make uneasy, abash," from Anglo-French *confoundre*, Old French *confondre* (12c.) "crush, ruin, disgrace, throw into dosorder," from Latin *confundere* "to confuse," literally "to pour together, mix, mingle," from assimilated form of *com* "together"+ *fundere* "to pour." http://www.etymonline.com/index.php?term=confound&allowed_in_frame=0.
§ Bias 1520s, from French *biais* "slant, slope, oblique," also figuratively, "expedient, means" (13c., originally in Old French a past participle adjective, "sideways, askance, against the grain"). http://www.etymonline.com/index.php?term=bias&allowed_in_frame=0.

relevant because there are other sources for confusion bias after random-ization (Morabia 2011; Vandenbroucke 2002). Therefore, if two philosophical issues support the obligation of the randomization process of clinical tri-als, why not respect them?! Why do they elude them?! If uncertainty is part of life, almost its essence, medicine with its probabilistic nature does not escape it. Based on this, and assuming the principle of good faith, why per-form RCTs that will increase or keep uncertainty? Is it ethical or moral? Why does need exist to address treatment uncertainties? "Ignoring uncertainties about the effects of treatments has led to avoidable suffering and deaths. To reduce this suffering and premature mortality, treatment uncertainties must be acknowledged and addressed, first by reviewing systematically what is already known, and then in well designed research to reduce continuing uncertainties" (The James Lind Library 2017). The research only seeks to find the truth, not to get truthlikeness* or verisimilitudes.[†] Poorly designed tri-als deviate from the lane leading to the final station: the truth. They travel on a path called "bias" that culminates in two stations: negative falsehoods or positive falsehoods. Being false negative yields damages and being false positive yields catastrophes. We must remember that design wrongs are not amended by statistics.

2.5 What Type of Randomization Process Are You Looking for?

When a clinical trial is truly randomized, the likelihood of getting infected by a bacteria called "allocation bias" is null. It is because proper randomization is a "vaccine" that has a number-needed-to-treat equal to 1. That bias is "an error in the estimate of an effect caused by failure to implement valid proce-dures for random allocation of subjects to intervention and control groups in a clinical trial" (Porta 2014). The mechanisms and direction of allocation bias in randomized clinical trials have recently been examined (Paludan-Müller et al. 2016). Potential allocation bias will seriously affect the findings and con-clusions of a systematic review (Higgins et al. 2011; Hróbjartsson et al. 2013; Jørgensen et al. 2016). An inappropriate RCT generates noise, which results in not only a presumed benefit but also a potential danger to the patient. On the contrary, adequate randomization of a clinical trial would reduce, among other things, a "slippage" in the size of the intervention effect estimator.

---

* Oddie, Graham, "Truthlikeness," *The Stanford Encyclopedia of Philosophy* (Winter 2016 Edition), Edward N. Zalta (ed.), URL = https://plato.stanford.edu/archives/win2016/entries/truthlike ness/. (Accessed 25 January 2017.)

† Verisimilitude, "appearance of thruth or reality, likelihood", c1600. From French *verisimilitude* (1540s), from Latin *verisimilitudo* "likeness to truth", from *veri*, genitive of verum, neuter of verus "true" + similis "like, similar" http://www.etymonline.com/index.php?term=veris imilitude&allowed_in_frame=0. (Accessed 25 January 2017.)

When a clinical trial is invaded by allocation bias "bacteria" unfailingly, that trial will develop two major complications: selection and confusion biases. The uncertainty will have prevailed!

The research team must use a method of assigning treatment against the accusation of selection bias (Gore 1981). It is a "bias in the estimated association or effect of an exposure on an outcome that arises from the procedures used to select individuals into the study or the analysis" (Porta 2014).

There are methods that assign (but not randomly) participants to the comparison groups (Egbewale 2014; Higgins et al. 2011) *Warning*! It is chance and not the choice of the researcher (Schulz and Grimes 2002). A clinical trial using strategies that do not guarantee the randomized allocation of participants to the comparison groups will be neither randomized nor ethical (Dunford 1990). Beware! Not all papers being introduced as randomized are randomized: Mr. Clinical Trial, what technique was used to perform the randomization for you? Clinical trials seldom report about how they guarantee randomness. The following approaches are improper randomization processes: alternately allocated (Edwards et al. 2016; Garg 2016; Okany 2004; Serjeant et al. 1970), odd-numbered/even-numbered (Eskola et al. 1990; Pond et al. 1995; Saad Setta et al. 2011), date of operations (Ahn et al. 2016), month of birth (Rizzolo et al. 1993; Rosenberg et al. 1975), and hospital number (More et al. 1985; Myerowitz et al. 1977; Yang et al. 2014). Obviously, these trials may have been controlled but not randomized. These papers were conducted with participants allocated by a nonrandomized process. Their randomness was known by the researcher team; thus, randomization had been predictable a priori. Look back on "Something is random if it happens by chance" (Eagle 2016). Why do research workers use illegitimate methods for randomizing? Who authorized performing these clinical trials? Who is responsible? This issue increases the likelihood of waste biomedical research. "Inadequate allocation concealment may exaggerate treatment effects in some trials while underestimate effects in others" (Paludan-Müller et al. 2016). It is very often that one reads papers labeled as randomized without any description of the method of randomization (Cofrancesco et al. 1994; Henk and Smith 1977; Ma et al. 2015; Richard et al. 1995; Saldalamacchia et al. 2004; Singla et al. 2014; Wang et al. 2002; Wun et al. 2013).

2.6 Is There an Association between Confusion Bias and Inappropriate Randomization Process?

It has been pointed out that "confounding" is a basic dilemma of comparability (Vandenbroucke 2002). Let us see. The randomization process has been dubbed "goddess" (Bangdiwala 2011). It has been assumed that randomization "blesses" a clinical experiment twice. The first "blessing" ensures that the participants in the clinical trial have the same chance to be assigned to any of the comparison groups (i.e., it is haphazard), and nothing more than chance decides the comparison group the participant will be assigned to. The second "blessing" is that the potential confounders or prognostic variables either known or unknown are evenly distributed among the comparison groups. However, that blessing is only fulfilled directly if the sample size is large! (Bangdiwala 2011). The researcher should be aware of the options for balancing these prognostic factors (Bangdiwala 2011; Green and Byar 1978; Kernan et al. 1999; Kraemer and Fendt 1990; Lachin 1988; Lin et al. 2015; Rigdon and Hudgens 2015). Potential imbalance of potential prognostic factors will wreak havoc to the statistical inference and will hinder the clinical decision-making (Feinstein and Landis 1976). Would such procedures work in clinical trials realized with 10 (Cofrancesco et al. 1994), 12 (Avvisati et al. 1989), 14 (Saldalamacchia et al. 2004), or 17 participants (Richard et al. 1995)? Will these studies be reproducible? Are they reliable? What is the degree of certainty? Do these results bear the weight of new research that seeks to corroborate those findings? It recalls Popper's theory: critical rationalism or falsificationism.* Table 2.1 illustrates the impact of the small sample size on the balance of the comparison groups in an RCT.

Table 2.1 shows four different random sequence generations of the potential RCT whose sample size was 17 participants (The James Lind Library 2017). We can observe that the first "blessing" of the "goddess" randomization resulted in different allocations of participants, while the second "blessing" of the "goddess" randomization yielded different distributions of potential known confusing variables. This will bring four types of statistical inferences, despite having the same intervention and the same type of control for a sample size of 17 participants. This results in inconsistency, heterogeneity, incoherence for clinical decision-making … and everything has been a product of allocation bias. It generates disorganization, which should be understood as non-comparability: confusion bias! If participants were randomized, then why are they not comparable? Subjects could be allocated by random, but that does not imply that the groups are comparable regarding baseline characteristics (Altman 1985; Lin et al. 2015). How to explain this lack of comparability? Let us take a look at Table 2.2.

As can be seen in Table 2.2, as the sample size increases, the comparison groups begin to be balanced. Thus, known prognostic factors are equally distributed, assuming the same for unknown prognostic factors. All this increases the certainty of statistical inference with the reduction of allocation bias: Fisher's advice! Of course, it will have a clear reduction of selection bias due to randomization. Observing Table 2.3, we will encounter another scene.

---

* Andersen, Hanne and Hepburn, Brian, "Scientific Method," *The Stanford Encyclopedia of Philosophy* (Summer 2016 Edition), Edward N. Zalta (ed.), URL = https://plato.stanford.edu /archives/sum2016/entries/scientific-method/. (Accessed 25 January 2017.)

TABLE 2.1

Unbalance of Comparison Groups in a Randomized Clinical Trial with Small Sample Size

17 Patients were Randomized, 12 to Treatment A, 5 to Treatment B:		
	Treatment A (total = 12) Number (%)	**Treatment B (total = 5) Number (%)**
Duration of Health Problem		
Long term	1(8)	1(20)
More recent	11(92)	4(80)
Severity of Health Problem		
Mild	5(42)	3(60)
Moderate	7(58)	1(20)
Severe	0(0)	1(20)
Age, in years		
Under 15	0(0)	1(20)
15–34	4(33)	2(40)
35–64	3(25)	0(0)
65 & older	5(42)	2(40)
Condition: Very Anxious?		
Yes	1(8)	0(0)
No	11(92)	5(100)
Make 10 times bigger		

17 Patients were Randomized, 7 to Treatment A, 10 to Treatment B:		
	Treatment A (total = 7) Number (%)	**Treatment B (total = 10) Number (%)**
Duration of Health Problem		
Long term	3(43)	4(40)
More recent	4(57)	6(60)
Severity of Health Problem		
Mild	0(0)	3(30)
Moderate	3(43)	4(40)
Severe	4(57)	3(30)
Age, in years		
Under 15	1(14)	0(0)
15–34	2(29)	4(40)
35–64	1(14)	3(30)
65 & older	3(23)	3(30)
Condition: Very Anxious?		
Yes	2(29)	0(0)
No	5(71)	10(100)
Make 10 times bigger		

(*Continued*)

TABLE 2.1 (CONTINUED)

Unbalance of Comparison Groups in a Randomized Clinical Trial
with Small Sample Size

17 Patients were Randomized, 11 to Treatment A, 6 to Treatment B:

	Treatment A (total = 11) Number (%)	Treatment B (total = 6) Number (%)
Duration of Health Problem		
Long term	2(18)	5(83)
More recent	9(82)	1(17)
Severity of Health Problem		
Mild	4(36)	1(17)
Moderate	4(36)	4(67)
Severe	3(27)	1(17)
Age, in years		
Under 15	2(18)	1(17)
15–34	3(27)	1(17)
35–64	3(27)	0(0)
65 & older	3(27)	4(67)
Condition: Very Anxious?		
Yes	0(0)	0(0)
No	11(100)	6(100)
	Make 10 times bigger	

17 Patients were Randomized, 10 to Treatment A, 7 to Treatment B:

	Treatment A (total = 10) Number (%)	Treatment B (total = 7) Number (%)
Duration of Health Problem		
Long term	3(30)	2(29)
More recent	7(70)	5(71)
Severity of Health Problem		
Mild	3(30)	2(29)
Moderate	4(40)	1(14)
Severe	4(40)	4(57)
Age, in years		
Under 15	3(30)	2(29)
15–34	2(20)	1(14)
35–64	2(20)	0(0)
65 & older	3(30)	4(57)
Condition: Very Anxious?		
Yes	1(10)	1(14)
No	9(90)	6(86)
	Make 10 times bigger	

Source: jameslindlibrary.org

TABLE 2.2

Impact of Sample Size Increasing on Balance of Comparison Groups in a Randomized Clinical Trial

17 Patients were Randomized, 10 to Treatment A, 7 to Treatment B:		
	Treatment A (total = 10) Number (%)	**Treatment B (total = 7) Number (%)**
Duration of Health Problem		
Long term	5(50)	0(0)
More recent	5(50)	7(100)
Severity of Health Problem		
Mild	3(30)	3(43)
Moderate	3(30)	2(29)
Severe	4(40)	2(29)
Age, in Years		
Under 15	3(30)	3(43)
15–34	3(30)	2(29)
35–64	1(10)	1(14)
65 & older	3(30)	1(14)
Condition: Very Anxious?		
Yes	0(0)	1(14)
No	10(100)	6(86)
	Make 10 times bigger	

17 Patients were Randomized, 10 to Treatment A, 7 to Treatment B:

170 Patients were Randomized, 85 to Treatment A, 85 to Treatment B				
	Treatment A (total = 10) Number (%)	**(total = 85)**	**Treatment B (total = 7) Number (%)**	**(total = 85)**
Duration of Health Problem				
Long term	5(50)	34(40)	0(0)	23(27)
More recent	5(50)	51(60)	7(100)	62(73)
Severity of Health Problem				
Mild	3(30)	21(25)	3(43)	18(21)
Moderate	3(30)	42(49)	2(29)	37(44)
Severe	4(40)	22(26)	2(29)	30(35)
Age, in Years				
Under 15	3(30)	26(31)	3(43)	14(16)
15–34	3(30)	22(26)	2(29)	27(32)
35–64	1(10)	8(9)	1(14)	19(22)
65 & older	3(30)	29(34)	1(14)	25(29)
Condition: Very Anxious?				
Yes	0(0)	4(5)	1(14)	6(7)
No	10(100)	81(95)	6(86)	79(93)
	Make 10 times bigger			

(*Continued*)

TABLE 2.2 (CONTINUED)

Impact of Sample Size Increasing on Balance of Comparison Groups
in a Randomized Clinical Trial

17 Patients were Randomized, 10 to Treatment A, 7 to Treatment B:

170 Patients were Randomized, 828 to Treatment A, 872 to Treatment B

	Treatment A (total = 10) Number (%)	(total = 828)	Treatment B (total = 7) Number (%)	(total = 872)
Duration of Health Problem				
Long term	5(50)	244(29)	0(0)	264(30)
More recent	5(50)	584(71)	7(100)	608(70)
Severity of Health Problem				
Mild	3(30)	255(31)	3(43)	276(32)
Moderate	3(30)	375(45)	2(29)	369(42)
Severe	4(40)	198(44)	2(29)	227(26)
Age, in Years				
Under 15	3(30)	167(20)	3(43)	167(19)
15–34	3(30)	206(25)	2(29)	229(26)
35–64	1(10)	206(25)	1(14)	199(23)
65 & older	3(30)	249(30)	1(14)	277(32)
Condition: Very Anxious?				
Yes	0(0)	45(5)	1(14)	51(6)
No	10(100)	783(95)	6(86)	821(94)
	Make 10 times bigger			

17 Patients were Randomized, 10 to Treatment A, 7 to Treatment B:

17,000 Patients were Randomized, 8540 to Treatment A, 8460 to Treatment B

	Treatment A (total = 10) Number (%)	(total = 8540)	Treatment B (total = 7) Number (%)	(total = 8460)
Duration of Health Problem				
Long term	5(50)	2514(29)	0(0)	264(30)
More recent	5(50)	6026(71)	7(100)	608(70)
Severity of Health Problem				
Mild	3(30)	2594(30)	3(43)	2438(29)
Moderate	3(30)	3858(45)	2(29)	3824(45)
Severe	4(40)	2088(24)	2(29)	2198(26)
Age, in Years				
Under 15	3(30)	1702(20)	3(43)	1761(21)
15–34	3(30)	2099(55)	2(29)	2182(26)
35–64	1(10)	2090(24)	1(14)	2041(24)
65 & older	3(30)	2649(31)	1(14)	2476(29)
Condition: Very Anxious?				
Yes	0(0)	413(5)	1(14)	441(5)
No	10(100)	8127(95)	6(86)	8019(95)
	Make 10 times bigger			

Source: jameslindlibrary.org

TABLE 2.3

Impact of High Sample Size on Balance of Comparison Groups at the Start of a Randomized Clinical Trial

1000 Patients were Randomized, 508 to Treatment A, 492 to Treatment B:

	Treatment A (total = 508) number (%)	Treatment B (total = 492) number (%)
Duration of Health Problem		
Long term	151 (30)	143 (29)
More recent	357 (70)	349 (71)
Severity of Health Problem		
Mild	150 (30)	142 (29)
Moderate	221 (44)	232 (47)
Severe	137 (27)	118 (24)
Age, in Years		
Under 15	102 (20)	94 (9)
15–34	138 (27)	131 (27)
35–64	121 (24)	122 (25)
65 & older	147 (29)	145 (29)
Condition: Very Anxious		
Yes	30 (6)	40 (8)
No	478 (94)	452 (92)
	Make 10 times bigger	

1000 Patients were Randomized, 508 to Treatment A, 492 to Treatment B:

100,000 Patients were Randomized, 5037 to Treatment A, 4963 to Treatment B:

	Treatment A (total = 508) number (%)	(Total 5037)	Treatment B (total = 492) number (%)	(Total 4963)
Duration of Health Problem				
Long term	151 (30)	1501 (30)	143 (29)	1487 (30)
More recent	357 (70)	3536 (70)	349 (71)	3476 (70)
Severity of Health Problem				
Mild	150 (30)	1492 (30)	142 (29)	1473 (30)
Moderate	221 (44)	2252 (44)	232 (47)	2254 (45)
Severe	137 (27)	1313 (26)	118 (24)	1236 (25)
Age, in Years				
Under 15	102 (20)	1017 (70)	94 (9)	987 (20)
15–34	138 (27)	1299 (26)	131 (27)	1241 (25)
35–64	121 (24)	1228 (24)	122 (25)	1256 (25)
65 & older	147 (29)	1493 (30)	145 (29)	1479 (30)
Condition: Very Anxious				
Yes	30 (6)	241 (5)	40 (8)	270 (5)
No	478 (94)	4796 (95)	452 (92)	4693 (95)
	Make 10 times bigger			

(*Continued*)

TABLE 2.3 (CONTINUED)

Impact of High Sample Size on Balance of Comparison Groups at the Start of a Randomized Clinical Trial

	1000 Patients were Randomized, 508 to Treatment A, 492 to Treatment B:			
	100,000 Patients were Randomized, 50,136 to Treatment A, 49,864 to Treatment B:			
	Treatment A (total = 508) number (%)	**(Total 50,136)**	**Treatment B (total = 492) number (%)**	**(Total 49,864)**
Duration of Health Problem				
Long term	151 (30)	14950 (30)	143 (29)	15019 (30)
More recent	357 (70)	35186 (70)	349 (71)	34845 (70)
Severity of Health Problem				
Mild	150 (30)	15116 (30)	142 (29)	14844 (30)
Moderate	221 (44)	22233 (44)	232 (47)	22619 (45)
Severe	137 (27)	12787 (26)	118 (24)	12401 (25)
Age, in Years				
Under 15	102 (20)	10007 (20)	94 (9)	9724 (20)
15–34	138 (27)	12715 (25)	131 (27)	12603 (25)
35–64	121 (24)	12612 (25)	122 (25)	12614 (25)
65 & older	147 (29)	14802 (30)	145 (29)	14923 (30)
Condition: Very Anxious				
Yes	30 (6)	2184 (5)	40 (8)	2584 (5)
No	478 (94)	47652 (95)	452 (92)	47280 (95)
		Make 10 times bigger		

Source: jameslindlibrary.org

Since its inception, an RCT with a sample size of 1000 participants has comparability in the groups (Table 2.3), such as it has been recently pointed out (Nguyen et al. 2017).

Summarizing, the randomization process not only assigns participants into comparison groups based on the sample size but also provides adequate balance in terms of known and unknown prognostic factors, such as comparability, to the comparison groups.

Should the editor-in-chief of any journal accept a paper labeled as "randomized" despite reporting a clear strategy of nonrandomization?

2.7 How Should the Randomization Section Be Reported?

Trial authors should use recommendations suggested by Standard Protocol Items: Recommendations for Interventional Trials (SPIRIT) statement (Chan et al. 2013). These suggestions are described in a 33-item checklist (Chan et al. 2013).

The following are examples of how the randomization process should be reported:

1. "A statistician blinded to the objectives and exact design of this study will use SAS 9.3 to generate the allocation sequence using block randomization (block size = 6). Researchers will take care so that block sizes do not become known to the researchers who conduct participant recruitment or intervention allocation. The researcher conducting participant recruitment at each site (Korean medicine doctor (KMD)) will contact the main trial site for central randomization by phone for random allocation to the three groups at a ratio of 1:1:1 of participants who have voluntarily agreed to participate in the trial, given written informed consent, and been screened and found eligible for trial participation in accordance with inclusion and exclusion criteria. Central randomization by phone will be performed through contact upon participant enrollment by the researcher conducting participant recruitment at each site to a designated independent researcher at the main trial site who will consecutively assign the randomized number and allocation group of each participant by verifying the sequence in the order of phone contact. All allocation numbers will be concealed through blinding of randomized number generation and central phone randomization. The four sites will receive random allocation competitively on participant enrollment without stratification by site, and each site will respectively stop recruitment when the equal allocated number of 15 participants has been recruited" (Shin et al. 2017).

2. "A randomisation list will be generated by an independent research assistant (not involved in data collection) using a computer random number generator. Allocation will be concealed in sequentially numbered sealed opaque envelopes. Given the influence of gender and foot strike pattern on lower limb mechanics during running, randomisation will be stratified to ensure balance of the treatment groups with respect to these variables (male/female; rearfoot strikers/non-rearfoot strikers). A block randomisation will also be used to make sure that three equal groups of 23 runners are obtained" (Esculier et al. 2016).

3. "After enrolment, patients will be randomised to either the EPVent group or the control group using a block randomisation scheme at a 1:1 ratio. Randomisation will be stratified by enrolment site. Assignment schedules will be generated for each clinical site, and each patient will be assigned using a web-based system (Studymaker. com) managed by the MCC" (Fish et al. 2014).

4. "Randomization was performed centrally on the basis of a block design, with stratification according to the number of crises in the preceding year (2 to 4 or 5 to 10) and concomitant hydroxyurea use (yes or no). Patients were assigned, in a 1:1:1 ratio, by an interactive Web- or voice-response system to receive low-dose crizanlizumab (2.5 mg per kilogram of body weight), high-dose crizanlizumab (5.0 mg per kilogram), or placebo" (Ataga et al. 2017).

As we can see, the extention report varies. Basically, five items should be reported:

1. The randomization technique
2. Reasons for which the randomization technique was chosen
3. The device used to perform the allocation sequence generation
4. How the sequence (randomness) was guaranteed
5. The known prognostic variables to stratify randomization

These items are the crux of a proper randomization. Transparency is the word!
These are clear examples of inappropriate reports on how to proceed with randomization:

1. "The ROMPA study is a multicentric, randomised, prospective, open clinical trial with a 28-day and 90-day follow-up and allocation ratio 1:1, assessing the mortality reduction by CPFA in patients with septic shock." "Patients will be divided randomly into two arms (control and intervention). ROMPA has a stratified randomisation based on gender, age ($\leq$65 or >65 years) and Simplified Acute Physiology Score (SAPS) III score (<50 or $\geq$51)" (Colomina-Climent et al. 2016).

2. "All subjects were selected 6 weeks after suffering a first ischemic stroke and randomized into parallel arms" (Alvarez-Sabín et al. 2016). The trial authors did not describe in the text how the randomizaton process was carried out!

2.8 Colophon

Since RCT has been considered as the "gold standard" for judging whether a treatment does more good than harm (Sackett et al. 1996), then it will be stripped of its "gold standard crown" status if it had been conducted with an inappropriate randomization process. If according to Bacon (1902),

"Numberless in short are the ways, and sometimes imperceptible, in which the affections colour and infect the understanding," why not use a proper randomization process to reduce the chance of waste biomedical research?

References

Adaptive Randomization. https://biostatistics.mdanderson.org/SoftwareDownload/SingleSoftware.aspx?Software_Id=62. Accessed 15th February 2017.

Ahn DK, Kim JH, Chang BK, Lee JI. Can we prevent a postoperative spinal epidural hematoma by using larger diameter suction drains? *Clin Orthop Surg* 2016;8(1): 78–83.

Altman DG. Comparability of randomised groups. *Journal of the Royal Statistical Society. Series D (The Statistician)* 1985;34(1): 125–136.

Alvarez-Sabín J, Santamarina E, Maisterra O, Jacas C, Molina C, Quintana M. Long-term treatment with citicoline prevents cognitive decline and predicts a better quality of life after a first ischemic stroke. *Int J Mol Sci* 2016;17(3): 390.

Ataga KI, Kutlar A, Kanter J, Liles D, Cancado R, Friedrisch J, Guthrie TH et al. Crizanlizumab for the prevention of pain crises in sickle cell disease. *N Engl J Med* 2017 Feb 2;376(5): 429–439.

Avvisati G, ten Cate JW, Büller HR, Mandelli F. Tranexamic acid for control of haemorrhage in acute promyelocytic leukaemia. *Lancet* 1989;2(8655): 122–124.

Bacon F. Novum Organum, by Lord Bacon, ed. By Joseph Devey, M.A. (New York: P.F.Collier, 1902). http://oll.libertyfund.org/titles/1432. Accessed 14/2/2017.

Bangdiwala SI. The 'goddess of randomisation' in small samples. *Int J Inj Contr Saf Promot* 2011;18(1): 89–93.

Berger VW. A review of methods for ensuring the comparability of comparison groups in randomized clinical trials. *Rev Recent Clin Trials* 2006;1(1): 81–86.

Berger VW, Bears JD. When can a clinical trial be called "randomized"? *Vaccine* 2003;21: 468–472.

Cassels A. *The Cochrane Collaboration: Medicine's Best-Kept Secret*. Canada: Agio Publishing House, 2015.

Chan A-W, Tetzlaff JM, Altman DG, Laupacis A, Gøtzsche PC, Krleža-Jerić K, Hróbjartsson A et al. SPIRIT 2013 Statement: Defining standard protocol items for clinical trials. *Ann Intern Med* 2013;158: 200–207.

Cipriani A, Nose M, Barbui C. Allocation concealment and blinding in clinical trials *Epidemiologia e Psichiatria Sociale* 2008;17, 2.

Cofrancesco E, Boschetti C, Leonardi P, Gianese F, Cortellaro M. Dermatan sulphate for the treatment of disseminated intravascular coagulation (DIC) in acute leukemia: A randomised, heparin-controlled pilot study. *Thrombosis Research* 1994;74(1): 65–75.

Colomina-Climent F, Giménez-Esparza C, Portillo-Requena C, Allegue-Gallego JM, Galindo-Martínez M, Mollà-Jiménez C, Antón-Pascual JL et al. Mortality reduction in septic shock by plasma adsorption (ROMPA): A protocol for a randomised clinical trial. *BMJ Open* 2016 Jul 12;6(7): e011856.

Dávalos A, Alvarez-Sabín J, Castillo J, Díez-Tejedor E, Ferro J, Martínez-Vila E, Serena J et al. Citicoline in the treatment of acute ischaemic stroke: An international, randomised, multicentre, placebo-controlled study (ICTUS trial). *Lancet* 2012;380(9839): 349–357.

Dunford FW. Random assignment: Practical considerations from field experiments. *Evaluation and Program Planning* 1990;13: 125–132.

Eagle A. "Chance versus Randomness," *The Stanford Encyclopedia of Philosophy* (Winter 2016 Edition), Edward N. Zalta (ed.), URL = https://plato.stanford .edu/archives/win2016/entries/chance-randomness/.

Eagle A. Randomness is unpredictability. *Brit J Phil Sci* 2005;56: 749–790.

Edwards L, Salisbury C, Horspool K, Foster A, Garner K, Montgomery AA. Increasing follow-up questionnaire response rates in a randomized controlled trial of telehealth for depression: Three embedded controlled studies. *Trials* 2016;17(1): 107.

Egbewale BE. Random allocation in controlled clinical trials: A review. *J Pharm Pharm Sci* 2014;17(2): 248–253.

Esculier JF, Bouyer LJ, Dubois B, Frémont P, Moore L, Roy JS. Effects of rehabilitation approaches for runners with patellofemoral pain: Protocol of a randomised clinical trial addressing specific underlying mechanisms. *BMC Musculoskelet Disord* 2016;17: 5.

Eskola J, Käyhty H, Takala AK, Peltola H, Rönnberg PR, Kela E, Pekkanen E, McVerry PH, Mäkelä PH. A randomized, prospective field trial of a conjugate vaccine in the protection of infants and young children against invasive *Haemophilus influenzae* type b disease. *N Engl J Med* 1990;323(20): 1381–1387.

Feinstein AR, Landis JR. The role of prognostic stratification in preventing the bias permitted by random allocation of treatment. *J Chronic Dis* 1976;29(4): 277–284.

Fish E, Novack V, Banner-Goodspeed VM, Sarge T, Loring S, Talmor D. The Esophageal Pressure-Guided Ventilation 2 (EPVent2) trial protocol: A multicentre, randomised clinical trial of mechanical ventilation guided by transpulmonary pressure. *BMJ Open* 2014;4(9): e006356.

Fisher RA, *Statistical Methods for Research Workers*, 1925.

Fisher RA, *The Design of Experiments*, 1935.

Freedman B. Equipoise and the ethics of clinical research. *N Engl J Med* 1987;317(3): 141–145.

Garg S. Outcome of intra-operative injected platelet-rich plasma therapy during follicular unit extraction hair transplant: A prospective randomised study in forty patients. *J Cutan Aesthet Surg* 2016;9(3): 157–164.

Gore, SM. Assessing clinical trials—Why randomise? *Br Med J (Clin Res Ed)* 1981;282 (6280): 1958–1960.

Green SB, Byar DP. The effect of stratified randomization on size and power of statistical tests in clinical trials. *J Chronic Dis* 1978;31(6–7): 445–454.

Henk JM, Smith CW. Radiotherapy and hyperbaric oxygen in head and neck cancer. Interim report of second clinical trial. *Lancet* 1977;2(8029): 104–105.

Higgins JPT, Altman DG, Sterne JAC (editors). Chapter 8: Assessing risk of bias in included studies. In: Higgins JPT, Green S (editors). *Cochrane Handbook for Systematic Reviews of Interventions* Version 5.1.0 (updated March 2011). The Cochrane Collaboration, 2011. Available from www.cochrane-handbook.org.

Howick J. *The Philosophy of Evidence-Based Medicine*. Wiley-Blackwell, Chichester, 2011.

Hróbjartsson A, Boutron I, Turner L, Altman DG, Moher D. Assessing risk of bias in randomised clinical trials included in Cochrane Reviews: The why is easy, the how is a challenge. *Cochrane Database Syst Rev* 2013 Apr 30;(4): ED000058.

Ioannidis JP. Clinical trials: What a waste. *BMJ* 2014;349:g7089.

Jørgensen L, Paludan-Müller AS, Laursen DR, Savović J, Boutron I, Sterne JA, Higgins JP, Hróbjartsson A. Evaluation of the Cochrane tool for assessing risk of bias in randomized clinical trials: Overview of published comments and analysis of user practice in Cochrane and non-Cochrane reviews. *Syst Rev* 2016;5: 80.

Kelly T, "Evidence," *The Stanford Encyclopedia of Philosophy* (Winter 2016 Edition), Edward N. Zalta (ed.), URL = https://plato.stanford.edu/archives/win2016/entries/evidence/.

Kernan WN, Viscoli CM, Makuch RW, Brass LM, Horwitz RI. Stratified randomization for clinical trials. *J Clin Epidemiol* 1999;52(1): 19–26.

Kim J, Shin W. How to do random allocation (randomization). *Clin Orthop Surg* 2014;6(1): 103–109.

Kraemer HC, Fendt KH. Random assignment in clinical trials: Issues in planning (Infant Health and Development Program). *J Clin Epidemiol* 1990;43(11): 1157–1167.

Lachin JM. Statistical properties of randomization in clinical trials. *Control Clin Trials* 1988;9(4): 289–311.

Lilienfeld AM. The Fielding H. Garrison Lecture: Ceteris paribus: The evolution of the clinical trial. *Bull Hist Med* 1982;56(1): 1–18.

Lin Y, Zhu M, Su Z. The pursuit of balance: An overview of covariate-adaptive randomization techniques in clinical trials. *Contemp Clin Trials* 2015;45(Pt A): 21–25.

Ma C, Hernandez MA, Kirkpatrick VE, Liang LJ, Nouvong AL, Gordon II. Topical platelet-derived growth factor vs placebo therapy of diabetic foot ulcers offloaded with windowed casts: A randomized, controlled trial. *Wounds* 2015;27(4): 83–91.

Medical Research Council (1948). Streptomycin treatment of pulmonary tuberculosis: A Medical Research Council investigation. BMJ 2: 769–782.

Morabia A. History of the modern epidemiological concept of confounding. *J Epidemiol Community Health* 2011;65(4): 297–300.

More DG, Raper RF, Munro IA, Watson CJ, Boutagy JS, Shenfield GM. Randomized, prospective trial of cimetidine and ranitidine for control of intragastric pH in the critically ill. *Surgery* 1985;97(2): 215–224.

Myerowitz PD, Caswell K, Lindsay WG, Nicoloff DM. Antibiotic prophylaxis for open-heart surgery. *J Thorac Cardiovasc Surg* 1977;73(4): 625–629.

Nguyen TL, Collins GS, Lamy A, Devereaux PJ, Daurès JP, Landais P, Le Manach Y. Simple randomization did not protect against bias in smaller trials. *J Clin Epidemiol* 2017;84: 105–113.

Okany CC, Atimomo CE, Akinyanju OO. Efficacy of natural honey in the healing of leg ulcers in sickle cell anaemia. *The Nigerian Postgraduate Medical Journal* 2004;11(3): 179–181.

Paludan-Müller A, Teindl Laursen DR, Hróbjartsson A. Mechanisms and direction of allocation bias in randomised clinical trials. *BMC Med Res Methodol* 2016;16(1): 133.

Pond SM, Lewis-Driver DJ, Williams GM, Green AC, Stevenson NW. Gastric emptying in acute overdose: A prospective randomised controlled trial. *Med J Aust* 1995;163(7): 345–349.

Porta M. *Dictionary of Epidemiology*. 6th edition. New York: Oxford University Press, 2014.

Random Allocation Software. http://mahmoodsaghaei.tripod.com/Softwares/ran dalloc.html. Accessed 15th February 2017.

Random.org. https://www.random.org/sequences/. Accessed 15th February 2017.

Randomization.com. http://www.randomization.com/. Accessed 15th February 2017.

Research Randomizer. https://www.randomizer.org/. Accessed 15th February 2017.

Richard JL, Parer-Richard C, Daures JP, Clouet S, Vannereau D, Bringer J, Rodier M et al. Effect of topical basic fibroblast growth factor on the healing of chronic diabetic neuropathic ulcer of the foot: A pilot, randomized, double-blind, placebo-controlled study. *Diabetes Care* 1995;18: 64–69.

Rigdon J, Hudgens MG. Randomization inference for treatment effects on a binary outcome. *Stat Med* 2015;34(6): 924–935.

Rizzolo SJ, Piazza MR, Cotler JM, Hume EL, Cautilli G, O'Neill DK. The effect of torque pressure on halo pin complication rates. A randomized prospective study. *Spine (Phila Pa 1976)* 1993;18(15): 2163–2166.

Rosenberg IL, Evans M, Pollock AV. Prophylaxis of postoperative leg vine thrombosis by low dose subcutaneous heparin or peroperative calf muscle stimulation: A controlled clinical trial. *Br Med J* 1975;1(5959): 649–651.

Saad Setta H, Elshahat A, Elsherbiny K, Massoud K, Safe I. Platelet-rich plasma versus platelet-poor plasma in the management of chronic diabetic foot ulcers: A comparative study. *International Wound Journal* 2011;8(3): 307–312.

Sackett DL, Rosenberg WM, Gray JA, Haynes RB, Richardson WS. Evidence based medicine: What it is and what it isn't. *BMJ* 1996;312(7023): 71–72.

Saldalamacchia G, Lapice E, Cuomo V, De Feo E, D'Agostino E, Rivellese AA, Vaccaro O. A controlled study of the use of autologous platelet gel for the treatment of diabetic foot ulcers. *Nutrition Metabolism and Cardiovascular Diseases* 2004;14: 395–396.

Schulz KF, Grimes DA. Generation of allocation sequences in randomised trials: Chance, not choice. *Lancet* 2002;359(9305): 515–519.

Sealed Envelope. https://www.sealedenvelope.com/simple-randomiser/v1/lists. Accessed 15th February 2017.

Serjeant GR, Galloway RE, Gueri MC. Oral zinc sulphate in sickle-cell ulcers. *Lancet* 1970;2(7679): 891–892.

Shin BC, Kim MR, Cho JH, Jung JY, Kim KW, Lee JH, Nam K, Lee MH, Hwang EH et al. Comparative effectiveness and cost-effectiveness of Chuna manual therapy versus conventional usual care for nonacute low back pain: Study protocol for a pilot multicenter, pragmatic randomized controlled trial (pCRN study). *Trials* 2017;18(1): 26.

Singla S, Garg R, Kumar A, Gill C. Efficacy of topical application of beta urogastrone (recombinant human epidermal growth factor) in Wagner's Grade 1 and 2 diabetic foot ulcers: Comparative analysis of 50 patients. *Journal of Natural Science, Biology and Medicine* 2014;5(2): 273–277.

Stanford Encyclopedia of Philosophy. https://plato.stanford.edu. Accessed 11 January 2017.

The James Lind Library 1.1 Why treatment uncertainties should be addressed. 2017. (http://www.jameslindlibrary.org/essays/1-1-why-treatment-uncertainties -should-be-addressed/)

Vandenbroucke JP. The history of confounding. *Soz Praventivmed* 2002;47(4): 216–224.

Viera AJ, Bangdiwala SI. Eliminating bias in randomized controlled trials: Importance of allocation concealment and masking. *Fam Med* 2007;39(2): 132–137.

Wang SE, Lara PN Jr, Lee-Ow A, Reed J, Wang LR, Palmer P, Tuscano JM et al. Acetaminophen and diphenhydramine as premedication for platelet transfusions: A prospective randomized double-blind placebo-controlled trial. *American Journal of Hematology* 2002;70: 191–194.

Wendler D. "The Ethics of Clinical Research," *The Stanford Encyclopedia of Philosophy* (Winter 2016 Edition), Edward N. Zalta (ed.), URL = https://plato.stanford .edu/archives/win2016/entries/clinical-research/. Accessed on 15th February 2017.

White SJ, Freedman LS. Allocation of patients to treatment groups in a controlled clinical study. *Br J Cancer* 1978;37(5): 849–857.

Wun T, Soulieres D, Frelinger AL, Krishnamurti L, Novelli EM, Kutlar A, Ataga KI et al. A double-blind, randomized, multicenter phase 2 study of prasugrel versus placebo in adult patients with sickle cell disease. *J Hematol Oncol* 2013 Feb 17;6: 17. doi: 10.1186/1756-8722-6-17.

Yang H, Sun R, Chang Y, Fu Y, Li B, Qin B, Lu Y et al. [A multicenter randomized controlled trial of sufentanil for analgesia/sedation in patients in intensive care unit]. *Zhonghua Wei Zhong Bing Ji Jiu Yi Xue* 2014;26(2): 94–100.

3

Sympathetic Bias: A Neglected Source of Selection Bias

William C. Grant

CONTENTS

3.1 Introduction 29
3.2 Investigator Preferences for Bias and Behaviors to Create Bias 29
    3.2.1 Motivations for Bias 29
    3.2.2 Behaviors Leading to Bias 31
3.3 Preferences Regarding Predictability and Sample Size 32
3.4 Conclusions and Policy Insights 37
References 37

3.1 Introduction

The chapter is organized as follows. Section 3.2 explains how clinical trial investigators may be motivated by fame, fortune, sponsor sympathy, and/ or patient welfare. We then describe various bias-creating behaviors that are made possible by treatment predictability. Section 3.3 describes a framework for understanding sponsor and investigator preferences, which is useful for depicting sympathetic bias. Section 3.4 concludes and mentions some clinical trial policy implications regarding sympathetic bias.

3.2 Investigator Preferences for Bias and Behaviors to Create Bias

3.2.1 Motivations for Bias

To clarify payoffs for clinical trial players, we first summarize the variety of motivations for investigators to create bias. There are at least four potential sources of preference for bias on the part of investigators: (1) fortune, (2) fame, (3) approbation from experimental sponsors, and (4) sympathy for patient well-being. Certainly, not every investigator seeks to bias their experiment, but a randomized experiment should defend against the possibility that one or more of these motivations underlie investigator preferences. It is precisely because we cannot observe investigators' motivations that experiments should minimize treatment predictability (Berger 2005).

Although the practice of paying investigators to produce results that are favorable to the sponsor is prohibited, pecuniary incentives may nonetheless be important because investigators can develop long-term relationships with sponsors. An experiment with results favorable to the sponsor may make it more likely that the investigator from that experiment wins contracts from that sponsor for future experiments. Investigators may also share profits if an experiment leads to a successful commercialization of the treatment being studied (Kjaergard and Als-Nielsen 2002).

A related concern is the potential desire for an investigator to achieve fame or recognition from the results of the experiment. An investigator may seek recognition in the form of academic publication, disciplinary prestige, professional honors, and so on. These could be indirectly tied to material income (Siegfried and White 1973; Tuckman and Leahey 1975) or they could produce utility directly (Levy 1988). A recent high profile case of investigator fraud illustrates just how strong the motivation for fame can be (*The Economist* 2011), and numerous retractions of scientific publications indicate fame as the motivational culprit (Retraction Watch 2012).

Separately, investigators may derive some utility from the approval of an experimental sponsor even if their fame and fortune were held constant. Levy and Peart (2007) describe how an investigator may have a sympathetic attraction to an experimental sponsor, based on an innate desire for the sponsor's gratitude. "Approbation," as they call this motivation, may develop from repeated association between the investigator and sponsor, in a variant of what Adam Smith called "affection" (Peart and Levy 2005).

Finally, investigators may have concern for the well-being of study subjects. Clinical ethics require that the practice of randomly assigning treatments to patients is unacceptable unless the scientific community is indifferent between the treatments a priori. Although this condition, known as collective equipoise, may hold for the scientific community as a group, it may not hold for each individual scientific investigator. Frequently, a particular investigator's prior knowledge will hold one treatment as preferred over another treatment *for some patients,* even before the experiment has been concluded. Such prior knowledge may develop from knowledge of earlier-stage experimental results, from previous experience with similar kinds of treatments, or from any variety of sources of clinical intuition (Byrne and Thompson 2006). When an investigator perceives one treatment as more promising for some patients, selection of particular patients to receive particular treatments may lead to bias (Berger 2005). This investigator motivation is altruistic in terms of therapeutic caregiving, but it is subversive in terms of statistical science.

3.2.2 Behaviors Leading to Bias

Having examined the various possible motivations for bias above, we now describe the bias-creating behaviors that are made possible by treatment predictability. These include (1) investigator selection of patients, (2) investigator influence on patient compliance, (3) investigator testimony on outcomes, and (4) investigator influence on psychological expectations. The payoffs for investigators will depend on how successfully these tactics can be employed by predicting treatments.

The most common behavior leading to bias is the strategic selection of patients (Berger 2005). Selection of a particular type of patient can be accomplished in a variety of ways. A common practice is enrollment discretion (Chalmers et al. 1983), where each investigator is allowed to deny enrollment based on his own subjective determination of the patient's suitability for the study. Penston (2003) details how the exclusion criteria are frequently quite vague, allowing the investigator to exclude patients for reasons "not specified by the protocol but by the responsible physician." When the favored treatment is predicted, the investigator may deny enrollment until a patient with a strong prognosis or chance of response comes along. In other trials, the investigator need only manipulate the appointment schedules to match favored treatment predictions with preferred patient types. Yet another possibility is for the investigator to remove a patient from the study (following the unmasking of an unfavored treatment) and then re-enroll them when the favored treatment is predicted (Brauer 2004).

Separately, an investigator may influence patient compliance: if the favored treatment is predicted, then the investigator may provide extra encouragement for the patient to follow the treatment directions. Elements of compliance include taking the appropriate number of pills, timing the treatment as directed, and avoiding contraindicated behaviors such as alcohol consumption. Treatment compliance is commonly much less than perfect and varies widely from one patient to the next. As a result, one patient may exhibit a higher or lower treatment effect than another patient because of, respectively, better or worse compliance (Cramer and Spilker 1992; Efron and Feldman 1991; Lesaffre et al. 2003; Oakes et al. 1993). For example, skipping more pill doses may reduce pharmacological effectiveness. By counseling patients more intensely to follow directions when the favored treatment is predicted, investigators may create bias.

Another potential behavior arises when predictions affect how the investigator subjectively measures treatment response variables. For example, the investigator may be asked to rate the patient's quality of life before and after the intervention on a subjective scale of 1 (very poor quality) to 7 (very good quality). Even physical functions like breathing may be subjectively rated, such as asking an investigator to rate how severely labored is an asthma patient's breathing. When the favored treatment is predicted, an investigator may understate the quality of life or other health status before treatment begins, which would upwardly bias the effect of the favored treatment on the subjective response variable (Delucca and Rhagavarao 2000).

Finally, investigators may create bias by communicating their predictions to patients, which could cause various kinds of placebo effects. A placebo effect is said to occur if a patient's confidence in the value of treatment influences the patient's health outcome, separately from any pharmacological treatment effect. Recent research shows that patient expectations are related to the probability of receiving a "good" treatment (Malani 2006, 2008). Statistical bias is created if the effects driven by patient expectations cannot be distinguished from the effects driven by pharmacological response. Investigators may create this kind of bias by sharing treatment predictions with patients.

3.3 Preferences Regarding Predictability and Sample Size

Sympathetic bias can be understood by portraying the interaction of the sponsor of a trial and the investigator for that trial in terms of a strategic game. Each of these players has preferences over the possible trial designs that depend on (1) the predictability of the design and (2) the sample size planned for the trial. Assume that the sponsor ranks each possibility (each unique combination of predictability and sample size) according to its profitability. Sample size can be assumed to initially exhibit positive but diminishing marginal revenue and positive increasing marginal cost. At this stage of experimentation, profits are assumed to be unequivocally increasing in predictability. This second assumption is a critical difference between sponsor preferences and investigator preferences. The idea is that profit expectations lead the sponsor to be *willing to let ineffective drugs go forward; they just don't want to miss an effective drug.* Higher predictability of a clinical trial design enables these goals.

The sponsor preferences can be depicted graphically using isoprofit curves, as are shown in Figure 3.1. Each curve shows equally profitable combinations of predictability (on the vertical axis) and sample size (on the horizontal axis) for the sponsor. In this example, the single most profitable point is labeled S*, where there is maximum predictability and an intermediate sample size. Each isoprofit curve emanating outward from S* is some less profitable set of predictability sample size values, with curves closer to S* being more profitable than those further away. Note that each isoprofit curve has a down-sloping segment, then a critical point at the down-peak of the curve, followed by an up-sloping segment. The down-sloping segment shows that, for any given profit level, the sponsor does not want too small of a sample size. Both in terms of statistical power and market response to the trial results, there is some range where the sponsor is so profit-motivated

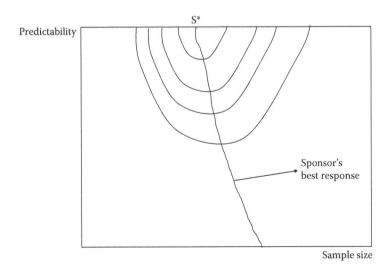

FIGURE 3.1
Isoprofit curves and best response function for the sponsor.

to achieve an adequately large sample size that they would actually be able to reduce predictability in order to increase the sample size and maintain expected profits. The up-sloping segment of each isoprofit curve corresponds to the range where sample size has become overly costly relative to the revenue benefits reflecting statistical power and expected market reception. In this up-sloping region of any isoprofit curve, the sponsor must be compensated with higher predictability if he is to maintain the same level of expected profits because the sample size has become unprofitably large. The down-peak of each isoprofit curve shows a special point. If we imagine that the sponsor chooses the sample size (since he holds the purse strings) and the investigator chooses the predictability (since she is the scientific expert who controls the clinical trial design), then each down-peak identifies the sponsor's best response to each possible predictability level chosen by the investigator. That is, conditional on the investigator's choice of predictability, the most profitable sample size possible for the sponsor corresponds to a down-peak of the sponsor's most profitable isoprofit curve. Connecting the locus of such down-peaks produces the sponsor's best response function, shown in Figure 3.1.

The investigator has her own set of iso-utility curves, which are not denominated in dollars of profit but are denominated in units of expected utility. Depending on whether or not the investigator is "sympathetic" to the sponsor, the investigator's iso-utility curves can take on a variety of shapes. Assuming that the investigator *is* sympathetic, her iso-utility curves are shown in Figure 3.2. Each investigator iso-utility curve has a negatively sloped portion, hits a critical point with infinitely vertical slope, and then

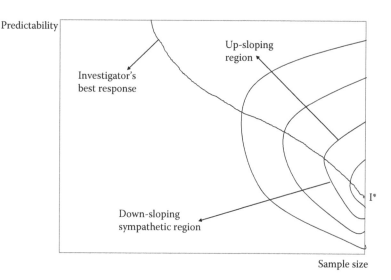

Predictability

Investigator's
best response

Up-sloping
region

Down-sloping
sympathetic region

I*

Sample size

FIGURE 3.2
Iso-utility curves and best response function for a sympathetic investigator.

bends forward into a positively sloped portion. The interpretations are very different from that of the sponsor. The entire mapping of investigator iso-utility curves is organized around the investigator most-preferred point, which is labeled I* on the right edge of the graph. As an investigator, higher sample size is more preferred because of the increase in statistical power and because the investigator does not have to spend her own dollars on the trial. However, because the investigator is sympathetic, the most preferred point I* does not involve zero predictability. Along the lines discussed in Section 3.2, the investigator actually prefers *some* greater-than-minimal level of predictability. Any outcome of the trial will result in some non-zero levels of type 1 and type 2 error, and the sympathetic investigator prefers some predictability so that it is less likely that an effective drug appears ineffective, even if the higher predictability also makes it more likely that an ineffective drug moves forward. The negatively sloped portion of each investigator's iso-utility curve is the sympathetic portion. Any reductions in sample size here must be compensated with higher predictability if the investigator is to maintain constant expected utility. The up-sloping portion is where the investigator's sympathy has essentially been critically depleted. All else equal, she would prefer that there is less predictability for any given level of sample size. Or, in other words, to keep expected investigator utility constant, increases in sample size must be offset by higher (more undesirable) predictability. Again, the critical point where each investigator iso-utility curve changes from negative slope to positive slope is special. If, again, we assume that the sponsor chooses the sample size and the investigator chooses the predictability, then the vertical coordinate of each side-peak equals the investigator's best response. For any given level of sample size imposed by the sponsor, the investigator's best choice of predictability (from a choice of trial design) corresponds to a side-peak of the investigator's iso-utility curve that can be achieved given that sample size. Connecting the side-peaks of all the investigator's iso-utility curves produces the investigator's best response function, which is identified in Figure 3.2.

The intersection of the sponsor and investigator best response functions is shown in Figure 3.3. The sponsor's choice of sample size and the investigator's choice of predictability can be portrayed as the Nash equilibrium shown in Figure 3.3. The extent of predictability and the sample size are each chosen by a player anticipating the other player's choice. In equilibrium, the predictability is greater than minimal because of the investigator's sympathy. Given the sample size determined by the sponsor, the investigator sympathetically makes the trial more predictable than necessary.

Most non-experts and, regrettably, many experts on clinical trials assume that the investigator's preferences are very different from the reality described above and depicted in Figure 3.2. A common false perception is that the investigator's expected utility is unconditionally decreasing in predictability, that is, that there is no sympathetic bias. To see the contrast, Figure 3.4 shows indifference curves for the fictional investigator motivated only by scientific validity. Predictability is "bad" in the language of microeconomics for such an unbiased investigator. The unbiased investigator's indifference curves are upward sloping throughout the entire predictability sample size space, indicating that higher predictability always reduces the investigator's expected utility for any given sample size. The most preferred

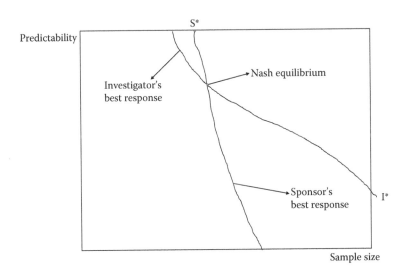

FIGURE 3.3
Nash equilibrium of sponsor sample size choice and investigator predictability choice.

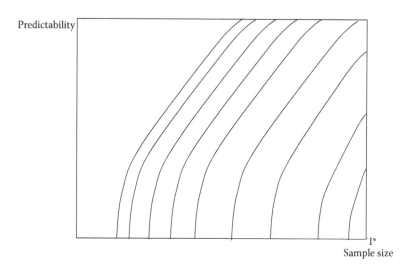

FIGURE 3.4
Fictional indifference curves for an unsympathetic investigator.

point on the graph (I*) is the most southeastern point where predictability is minimized and sample size is maximized. To maintain any given level of expected utility, the investigator must be compensated with a higher sample size to offset any increase in predictability from the trial's design. Each indifference curve lying further rightward (further downward) is more preferred.

The fictional investigator in Figure 3.4 would interact with a profit-motivated sponsor in ways leading to a very different Nash equilibrium than that depicted in Figure 3.3. The unsympathetic investigator's best response to *every* possible sample size is to choose a trial design with zero predictability. The fictional Nash equilibrium, which would result if investigators were actually unbiased, is shown in Figure 3.5 where the profit-motivated sponsor's best response function crosses the *x* axis.

This is the point where the sponsor is forced to accept zero predictability and just pay for the most profitable sample size given no predictability. In terms of scientific validity, this Nash equilibrium in Figure 3.5 would be a twofold gain: the sample size is bigger and the predictability is lower. When the possibility of excessive predictability is taken off the table (due to the preferences of an unsympathetic investigator), the sponsor prefers to shell out for a bigger trial in order to increase the chances of showing that he has an effective new product. The fictional Nash equilibrium (in Figure 3.5) drives the profits of the sponsor lower than the reality of sympathetic bias (shown in Figure 3.3), but still leaves the sponsor with some power to enhance expected profits by reducing the sample size below the level that would be more preferred by the investigator.

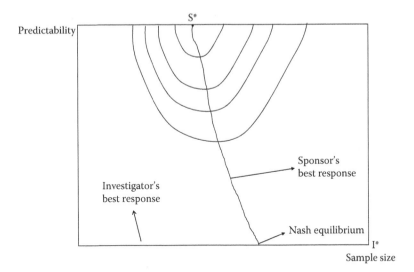

FIGURE 3.5
Fictional Nash equilibrium with an unsympathetic investigator.

3.4 Conclusions and Policy Insights

The issue of sympathetic bias is unavoidable. Investigators are human beings whose interactions with sponsors naturally lead to their sympathy and overlapping preferences when it comes to trial predictability. The biggest threat resulting from sympathetic bias comes from those who deny it exists. It is important to understand that the standard modes of operation when it comes to clinical trial design (especially method of randomization) are not without bias. The use of excessively predictable randomization methods (such as permuted block design) does not imply that these are the most valid methods, but rather that there is a natural self-interested inertia shared by trial sponsors and investigators. Clinical trials, like so many areas of human interaction, are situations where the self-perpetuating outcome is less than optimal. Acknowledging this reality is the first step toward making socially desirable improvements.

References

Berger, V.W., 2005. *Selection Bias and Covariate Imbalance in Randomized Clinical Trials.* John Wiley & Sons.

Brauer, C., 2004. Failure analysis of clinical studies with medical devices. Presentation at the Henry Stewart Talks Conference, Gaithersburg, MD, USA, October 27, 2004.

Byrne, M.M., Thompson, P., 2006. Collective equipoise, disappointment and the therapeutic misconception: On the consequences of selection for clinical research. *Medical Decision Making* 26(5), 467–479.

Chalmers, T.C., Celano, P., Sacks, H.S., Smith, H., 1983. Bias in treatment assignment in controlled clinical trials. *New England Journal of Medicine* 309, 1358–1361.

Cramer J., Spilker B. (eds.), 1992. *Non-Compliance in Clinical Trials*, Raven Press, New York.

Delucca, P., Rhagavarao, D., 2000. Effect of investigator bias on the significance level of the Wilcoxon rank-sum test. *Biostatistics* 1(1), 107–111.

Efron, B., Feldman, D., 1991. Compliance as an explanatory variable in clinical trials. *Journal of the American Statistical Association* 86, 9–26.

Kjaergard, L., Als-Nielsen, B., 2002. Association between competing interest and authors conclusions: Epidemiological study of randomized clinical trials published in the BMJ. *British Medical Journal* 325, 249.

Lesaffre, E., Kocmanova, D., Lemos, P.A. et al., 2003. A retrospective analysis of the effect of noncompliance on time to first major adverse cardiac event in LIPS. *Clinical Therapeutics* 20, 2431–2447.

Levy, D.M., 1988. The market for fame and fortune. *History of Political Economy* 20(4), 615–625.

Levy, D.M., Peart, S.J., 2007. Sympathetic bias. *Statistical Methods in Medical Research* 17, 265–277.

Malani A., 2006. Identifying placebo effects with data from clinical trials. *Journal of Political Economy* 114(2), 236–256.

Malani, A., 2008. Patient enrollment in medical trials: Selection bias in a randomized experiment. *Journal of Econometrics* 144(2), 341–351.

Oakes D., Moss A., Fleiss J. et al., 1993. Use of compliance measures in an analysis of the effect of diltiazem on mortality and reinfarction after myocardial infarction. *Journal of the American Statistical Association* 88(421), 44–49.

Peart, S.J., Levy, D.M., 2005. *The Vanity of the Philosopher: From Hierarchy to Equality in Post-classical Economics*. University of Michigan Press.

Penston, J., 2003. *Fiction and Fantasy in Medical Research: The Large-Scale Randomized Trial*. The London Press.

Retraction Watch, 2012. Retrieved February 1, 2012 from http://retractionwatch.wordpress.com/.

Siegfried, J.J., White, K.J., 1973. Financial rewards to research and teaching: A case study of academic economists. *American Economic Review* 63, 309–316.

The Economist, 2011. Misconduct in science: An array of errors. September 10–16, 2011, 91–92.

Tuckman, H.P., Leahey, J., 1975. What is an article worth? *Journal of Political Economy* 83, 951–968.

4

The Alleged Benefits of Unrestricted Randomization

Vance W. Berger

CONTENTS

4.1 Introduction 39
4.2 Unrestricted Randomization: Theory and Practice 40
4.3 Which MTI Value Governs the Selection Mechanism? 42
4.4 Which MTI Procedure Is Induced by the Selection Mechanism? 43
4.5 A Clarifying Dialogue 45
4.6 Conclusions 47
References 48

4.1 Introduction

Improper randomization, most notably permuted block randomization with its excessive set of restrictions, is known to compromise allocation concealment, thereby opening the door to a form of selection bias that can completely invalidate a trial. It is worth pointing out that, in general, a method that is vulnerable to bias is a flawed method, regardless of how often the bias actually occurs in practice. This is not to say that selection bias is uncommon. On the contrary, there are some suggestions that it occurs rather often. See, for example, Chapter 3 of Ref. [1], which documents 30 actual trials that are suspected of exactly this type of selection bias, and also makes clear that the detection probability is so low that these 30 almost certainly represent only the tip of the iceberg.

It is one thing to recognize an existing procedure as flawed, and quite another to find a better way. Don't discard your dirty water, as they say, until you find clean water. But in fact, we have found clean water, and more than one source of it. Specifically, a great deal of recent work has focused on the so-called MTI (constructed using a boundary based on the MTI, or maximally tolerated imbalance) randomization procedures (including the big stick [2], Chen [3], maximal [4], and block urn [5] procedures), as these MTI procedures are all substantially less restrictive, and therefore substantially more resistant to selection bias, compared to permuted block randomization.

In fact, these four MTI procedures all use the same set of restrictions on the randomization and differ only with respect to the conditional allocation probabilities when there is some imbalance that does not yet reach the MTI boundary (see Table II of Ref. [6]).

Given that only unrestricted randomization can completely eliminate the type of selection bias we consider [1], some authors point to the few trials that have used unrestricted randomization, and lament the fact that more do not follow suit. For example, Kahan et al. [7] found that only 3% of the trials they considered used unrestricted randomization, and they called for more trials to do the same in the future.

While the goal of eliminating selection bias is certainly a laudable one, we argue that the advice is misguided, that in fact 3% of the trials did *not* use unrestricted randomization, and that if they did, then that would be too many, not too few. The reality is that truly unrestricted randomization never has been, and never will be, used in an actual trial, nor could it ever be [8]. We shall comment on what is *actually* used when unrestricted randomization is claimed, and note that the resulting randomization procedure cannot be identified, but *may* in fact be the maximal procedure (with unknown MTI value). This consideration would give us yet another compelling reason to use the maximal procedure in future trials, a bit of lagniappe if you will, but of course with the caveat that researchers have to be upfront and explicit about this use.

4.2 Unrestricted Randomization: Theory and Practice

In theory, unrestricted randomization is quite simple. Every allocation sequence of the desired length is admissible, and we have only to select one at random, with equal probabilities for all. This is part and parcel of unrestricted randomization, the lack of any restrictions characterizing its identity. That is to say, unrestricted randomization is characterized by the fact that knowledge of some or even all of the past allocations will not offer any help in guessing the future ones. But let us compare and contrast two different variations on this theme, to highlight the point we wish to make.

Variation 1:

Enumerate all possible allocation sequences of the desired length.

Assign each of these allocation sequences the same probability.

Randomly select one.

Note that it is admissible.

Use it to randomly allocate the patients.

Compare that to the second variation, which is almost identical, and yet fundamentally different owing to what at first seems to be a slight, even semantic, change in wording.

Variation 2:

Enumerate all possible allocation sequences of the desired length.

Assign each of these allocation sequences the same probability.

Randomly select one.

Stipulate to accepting as admissible any sequence, no matter what it is.

Use it to randomly allocate the patients.

It is only the fourth step that differs, and this difference is subtle, yet quite important. Suppose that the trial is designed to randomize 100 patients to each of two treatment groups, so that the overall sample size is 200. The likelihood of drawing either of the two least favorable sequences, all 200 patients to the active treatment or all 200 patients to the control, is astronomically small. But it is not zero. It *could* happen. And if it did, then would we accept it? Or would we throw this one back, and try again? What about 199 patients allocated to one treatment group, and only 1 patient allocated to the other? Where exactly do we draw the line? If there is any line at all, then no matter where it is drawn, we are not using unrestricted randomization. The second variation is in fact unrestricted randomization, as it accepts the sequence no matter what it is. The first variation is the one used in practice, and it has embedded within it a mechanism for accepting or rejecting the allocation sequence that it obtains. And this, of course, begs the question of which allocation sequences would be rejected?

A moment's thought will reveal that what we are really after here is nothing more, or less, than the maximally tolerated imbalance (MTI) that is used explicitly in the construction of the MTI procedures. Clearly, no researchers would ever consent to a trial in which all patients receive the same treatment as a result of an unfortunate draw from the list of allocation sequences. Yet, failure to do so constitutes a violation of the terms of agreement implicit in the use of unrestricted randomization. As Rawls [9] notes, "when we enter an agreement we must be able to honor it even should the worst possibilities prove to be the case" (p. 176). The worst possibilities (from our perspective), all patients allocated to the same treatment group, would not in any way minimize our loss compared to the worst outcomes of other strategies we could choose, and in fact it might well even maximize the maximum loss.

So while decision theory suggests that we consider a minimax procedure, unrestricted randomization is, instead, a "maximax" procedure, and therefore cannot be recommended. However, the worst-case scenario, were it to arise, would never be honored anyway, so the agreement has been violated *a priori*, and unrestricted randomization has, in fact, *not* been used. This may sound like a good thing. We just criticized unrestricted randomization on the basis of its tendency to maximize the maximum loss, so not using it might appear to be just what we are looking for. Alas, what we are looking for, first and foremost, is truth in advertising. The identity of the specific procedure used when unrestricted randomization is claimed can never be inferred for certain on this basis alone. However, we can at least offer a fairly good guess as to what *has* been used in its place. We will discuss this inference in Section 4.3.

4.3 Which MTI Value Governs the Selection Mechanism?

The actual procedure used when unrestricted randomization is claimed is one that is filtered through an unspecified filter. That is to say, there is a subset of allocation sequences that are deemed acceptable (or would be if they came up), and all goes according to plan as long as one of these acceptable sequences is chosen. Admittedly, this is quite likely. However, it is not guaranteed, and one tenet of good research is to consider all possibilities and contingencies. So what happens if a sequence is selected and it is considered unacceptable? In this case, the remedy is rather simple. A new sequence is picked as if the first one never had been (no harm, no foul, as some might say), and this alternating selection and inspection or vetting process continues until a suitable sequence is found. This process will, in general, not require very many iterations, given the overwhelming likelihood that any sequence chosen will be admissible. But this vetting process is almost certainly conducted on an *ad hoc* basis, with inspection taking place only *after* the candidate sequence is chosen, as opposed to prospectively defining which sequences are admissible and which are not. This distinction is not mere semantics.

After the fact, even in the best situation (complete information being furnished), one can still only guess as to what the precise filtering mechanism was that governed the actual selection of the randomization procedure. It seems rather unlikely that complete information would be furnished, however, because acknowledging earlier outcomes of the randomization process that were not accepted would lead to questions that one would just as soon not face. Therefore, any effort at reconstruction is hindered by intentional censoring of what actually took place. Moreover, even if this information *is* provided, the data are still rather sparse, and can hardly be expected to provide an adequate basis for determining the acceptable and unacceptable subsets of allocation sequences. At best, we would be dealing with interval censoring, as we would know the sequence chosen (and its largest imbalance) as well as the levels of imbalance of the sequences rejected. Only if a sequence was rejected that reached a level of imbalance one greater than that of the chosen sequence can we assert that we know the value of the inferred or induced MTI (and even here, this is the case only if the latent MTI being used implicitly was stable, which need not be the case). Otherwise, we are limited to confining it to an interval.

However, we can at least offer an educated guess regarding the *nature* of the delineation between admissible and unacceptable, if not the delineation itself. We do not want too many patients going to one treatment group and too few going to the other. We also do not want a bad distribution over time, so that very many more early patients are allocated to one treatment group, and very many more late patients to the other one, for fear of chronological bias [10]. This equates to using an MTI, just as the MTI procedures do. So there are likely two distinct types of restriction operating. First, the terminal imbalance cannot be too large. Second, the largest imbalance over the course of the trial must also avoid getting too large.

Of course, what constitutes "too large" need not be the same, numerically, for the two restrictions. It is entirely reasonable to allow greater imbalance at some point during the trial than at the end of the trial, which may or may not be forced to balance perfectly. We see from this analysis that (1) trials claiming to use unrestricted randomization are actually instead using a version of it that is projected onto a smaller space, the space (subset) of allocation sequences that are deemed admissible, and (2) this admissible subset is almost certainly defined by adherence to an MTI condition for a suitable MTI value.

If the largest imbalance at any point in the sequence exceeds the MTI, then this sequence is rejected. One sequence is chosen at random from among those whose largest numerical imbalance never does exceed the MTI. We are projecting a larger space onto a smaller one defined by an MTI condition. That is to say, we know that a nontrivial MTI (meaning less than $2n$, the total number of patients) is used implicitly in the variation of unrestricted randomization that is used in practice. But since it is implicit, we do not know which MTI value is in effect. Chances are that the researchers themselves never actually confront this issue, and instead take the easy way out ... "I don't know precisely how much imbalance is too much, but I will know it when I see it." In other words, the MTI is derived on the fly, and this has negative consequences for trial validity.

4.4 Which MTI Procedure Is Induced by the Selection Mechanism?

As noted earlier, the big stick, Chen, maximal, and block urn procedures all use the same MTI condition, yet they are still different procedures by virtue of using different conditional allocation probabilities when there is some imbalance but not yet reaching the MTI [6]. We might wonder, then, which MTI procedure results from our compound procedure of conducting unrestricted allocation with a selection mechanism tacked on. Is it one that we have already considered, or is it instead an as-yet unnamed procedure? Intuitively, it might seem logical to suppose that we would end up with the big stick procedure, since this procedure is defined in such a way as to make it seem like the natural analog for unrestricted randomization. Specifically, one simply adds on the MTI condition, or big stick, to unrestricted random-ization. This entails that every allocation conducted when the imbalance does not reach the MTI value is essentially the flip of a fair coin, just as unre-stricted randomization is. The big stick procedure, therefore, differs from unrestricted randomization only when the MTI is reached, and may, on that basis, appear to be "closest" to unrestricted by any reasonable metric.

But consider that one facet of replacement randomization [1,4] is leaving the relative selection probabilities intact. Since unrestricted randomiza-tion specifies equal selection probabilities for each allocation sequence, our induced MTI procedure, whatever it is, must also abide by equal selection probabilities for each admissible sequence. The big stick procedure does not enjoy this property.

For our purposes, an allocation procedure is nothing more, or less, than a set of allocation sequences and assigned non-negative selection probabili-ties for each, with these probabilities summing to one. In other words, it is just a probability distribution over the set of allocation sequences. With this characterization in mind, we can define two randomization procedures to be aligned if (1) one set of allocation sequences is a subset of the other and (2) if the selection probabilities are proportional to each other for all sequences in the smaller set. That is, let A be a subset of B. Then, for each sequence X in A, we require that $P\{X|A\} = P\{X|B\}/P\{A|B\}$ be the probability of X being selected if we choose from among A only. This condition may appear to be satisfied as a tautology, and this owes to a slight abuse of notation, as the left-hand side, $P\{X|A\}$, can be used to mean either the conditional probability of sequence X, when we use allocation procedure B, given that we know that the chosen sequence belongs to subset A. It can also mean the unconditional probability of sequence X when we use allocation procedure A. The condition we require equates these two quantities for all sequences X in A. In general, these two expressions need not be equal.

As already noted, B in our case (unrestricted randomization) specifies equal probabilities for all sequences X in B. Among the MTI procedures, only the maximal procedure also specifies equal probabilities for all sequences (or at least for all sequences found to satisfy the MTI condition). Therefore, we see that when unrestricted randomization is projected onto the subset defined by the MTI condition, the maximal procedure results. In fact, it was suggested [4] that the maximal procedure can be implemented precisely in this manner, with replacement randomization used to discard inadmissible sequences, and sampling continuing until an admissible one is found. In Ref. [4], the sampling would be done from among all sequences with terminal balance (i.e., an equal number of allocations to each treatment group), but it can just as easily be done with unrestricted randomization (meaning, no requirement for terminal balance) as the base model generating the candidate sequences.

4.5 A Clarifying Dialogue

Given that unrestricted randomization has never been used, and will never be used, we find that a few issues confront us. What is a journal to do with an article that tallies up the frequencies with which each randomization method is used, and includes unrestricted randomization as a category? What is a journal to do with an article whose authors claim to have used unrestricted randomization? What is a journal to do with an article that proposes that unrestricted randomization be used more often? What is a regulatory agency to do with a trial that ostensibly used unrestricted randomization? What is a member of a research team to do if other members of this team propose unrestricted randomization for the trial being planned? In all of these cases, the answer is the same, which is to use the Socratic method, to question until the truth is found. The following sample dialogue between two hypothetical researchers, A and B, might be helpful.

A: I see that you are planning to use unrestricted randomization for the trial.
B: Yes, this is the only randomization method that can guarantee the complete elimination of selection bias, through complete allocation concealment.
A: Yes, that is true, and I certainly applaud your interest in eliminating selection bias and bolstering allocation concealment. I share this interest. But still, unrestricted randomization has some drawbacks, doesn't it?
B: Well, I suppose that it is not ideal for the control of chronological bias.
A: Yes, that is true. Even if we were to specify equal numbers of patients in each treatment group (which of course is contrary to the definition of unrestricted randomization, since terminal balance itself does constitute a restriction), we could still get very many more early allocations to one treatment group and a compensating number of late allocations to the other treatment group. This would confound treatments and time, and would result in baseline imbalances if there are any time trends.
B: But the laws of probability tell us that it is exceedingly unlikely that we will see these types of imbalance.
A: No, the laws of probability make no such claim. True, a symmetric random walk in one dimension is recurrent, so it will return to the

origin (perfect balance) infinitely often. But in no way does this rule out long trips away from the origin (recall the famous gambler's ruin problem). In fact, each level of imbalance will also be reached infinitely often, over the long run.

B: OK, so large imbalances are theoretically possible, but aren't they still rather unlikely? Do we really need to lose sleep over this?

A: I do not think that this outcome is as unlikely as it is generally believed to be, or as unlikely as *you* seem to believe it to be. But, moreover, how many pivotal trials of this medical intervention are you planning?

B: Ideally, we would like to conduct as many trials as possible, but the costs are prohibitive. Therefore, we are in a position to conduct just this one.

A: If this is your one shot at establishing definitive safety and efficacy, then is it really ideal to take chances when you do not have to?

B: Well, permuted block randomization would guard against this chronological bias you speak of, but at what price? We are then out of the frying pan and into the fire, since by protecting us against chronological bias it also renders us highly susceptible to selection bias. I argued strongly against permuted blocks when this came up in the meeting.

A: As well you should. It is ironic, is it not, that the most popular randomization method is also arguably the worst. But there are other (better) options too. Permuted blocks and unrestricted randomization are not your only options.

B: Those were the only ones I was aware of, other than minimization and alternation, neither of which qualifies as true randomization anyway. What are these better ones?

A: We will get back to that, but first, I would like to explore another problem with unrestricted randomization. Aside from chronological bias, aren't you aware that without specifying terminal balance unrestricted randomization can also result in grossly unbalanced treatment groups, in terms of group sizes? The active treatment group could end up with too few patients to even allow for an adequate safety profile.

B: Yes, I suppose that by conveniently putting this possibility out of my mind I was really taking my chances, wasn't I?

A: You were, but, as I mentioned before, there is a better way. Let me ask you what you would have done had it turned out that literally *all* the patients were allocated to the control group, and none to your active treatment.

B: That would never happen.

A: Still in denial, I see. True, it is rather unlikely, but it is one of the possible outcomes. What would you have done had this been the result of the randomization?

B: Then we would not even be studying our treatment at all, and we are not in a position to sponsor a study of the control.

A: So then, I take it you would have rejected this allocation sequence, and would have randomized again?

B: Of course. Wouldn't you?

A: Well, I would not have randomized in this way in the first place, but if I were to find myself in that situation, then yes, I would have to agree with you. But don't you see that this in and of itself already means that you did not actually mean it when you said that you wanted unrestricted randomization? You are restricting it, manually as it turns out, so as to exclude that one outcome, and, I would imagine, others too.

B: Well obviously. We are not going to sponsor a trial of the control.

A: Can you enumerate all allocation sequences that you would reject? Or at least describe them in some way that characterizes them?

B: That sounds like a lot of work.

A: It doesn't have to be. I suspect that the heart of the matter is the largest imbalance attained, in the relative treatment group sizes, over the duration of the trial. Would you agree with that?

B: Yes, that does sound reasonable.

A: How large an imbalance would you accept?

In this way, the implicit MTI value can be elicited.

4.6 Conclusions

Starr [11] (p. 47) commented on what he termed "cultural sedimentation," or the phenomenon by which "the theories and remedies of learned traditions filter down to the lower classes, where they remain even after the learned have abandoned them." He was *not* referring to randomization methods in randomized clinical trials, but he may as well have been, since this does describe the situation fairly well. In particular, the most primitive randomization methods were heralded as a great advance to the field for reasons that are well understood and need not be repeated here (but pertain to all the advantages randomized trials have over observational studies). But these primitive methods allowed for vastly unequal treatment groups, and also for chronological bias, so they were replaced with the next advance, permuted blocks, which, at least within a stratum, can guarantee that the group sizes will not differ by more than a specified amount. Alas, these methods also allow for selection bias, which is arguably worse.

The "learned" in this case recognize this shortcoming and have therefore turned away from permuted blocks, whereas the "lower" classes (at least with regard to how trials should be randomized) have steadfastly clung to what they know, what was once presented as, the sacred word of the learned who proclaimed that permuted block randomization is the way to go. But even among the relatively few sufficiently enlightened to recognize the drawbacks of permuted block randomization, there is still a dichotomy regarding the remedy. One group apparently reasons that unrestricted randomization was the best method before permuted blocks came along, and permuted blocks have been amply demonstrated to be fatally flawed, so therefore we go back to unrestricted randomization. The other group prefers to move forward, rather than backward, and has instead lauded the merits of MTI randomization.

There is no question that MTI randomization is vastly superior to either permuted block randomization or unrestricted randomization [6]; it is not a compromise, as some cynics might claim it is. Instead, it is the best of both worlds, controlling chronological bias as well as permuted block randomization does, while simultaneously controlling selection bias far better than blocked randomization does, and almost as well as unrestricted randomization does. Moreover, it also avoids the problem considered here of unsuitable allocation sequences, which can occur with unrestricted randomization. The only valid criticism of MTI randomization is that it fares worse than blocked randomization when all unmasking occurs after a long delay, but even this has been addressed in the literature [12].

At this point, it is rather clear, in light of the overwhelming evidence (which has never been contradicted), that MTI randomization is optimal. It is less clear-cut which specific MTI randomization method is best, but for a variety of reasons, one of which has been articulated in this chapter, the maximal procedure, or its newer cousin the asymptotic maximal procedure [13], seems best suited for actual trials. It is high time that one of these versions of the maximal procedure, or at least another MTI procedure, be used not only when it is incorrectly described as unrestricted randomization but also for all randomized trials.

References

1. Berger, VW (2005). *Selection Bias and Covariate Imbalances in Randomized Clinical Trials.* Chichester: John Wiley & Sons.
2. Soares, JF, Wu, CFJ (1983). Some restricted randomization rules in sequential designs, *Communications in Statistics Theory and Methods* 12, 2017–2034.
3. Chen, YP (1999). Biased coin design with imbalance tolerance. *Communications in Statistics* 15, 953–975.

4. Berger, VW, Ivanova, A, Deloria-Knoll, M (2003). Minimizing predictability while retaining balance through the use of less restrictive randomization procedures. *Statistics in Medicine* 22(19), 3017–3028.

5. Zhao, W, Weng, Y (2011). Block urn design—A new randomization algorithm for sequential trials with two or more treatments and balanced or unbalanced allocation. *Contemp Clin Trials* 2011 Nov; 32(6), 953–961.

6. Berger, VW, Agnor, RC, Bejleri, K (2016). Comparing MTI randomization procedures to blocked randomization. *Statistics in Medicine* 35(5), 685–694.

7. Kahan, BC, Rehal, S, Cro, S (2015). Risk of selection bias in randomised trials. *Trials*, in press.

8. Berger, VW (2017). Risk of selection bias in randomized trials: Further insight. *Trials* 17, 485–489.

9. Rawls, J (1971). *A Theory of Justice*. Cambridge: Belknap Press.

10. Matts, JP, McHugh, RB (1983). Conditional Markov chain designs for accrual clinical trials. *Biometrical Journal* 25, 563–577.

11. Starr, P (1982). *The Social Transformation of American Medicine*. New York: Basic Books.

12. Berger, VW, Agnor, RC (2016). Delayed unmasking and selection bias: Reply to crisp. *Statistics in Medicine* 35(12), 2111–2112.

13. Zhao, WL, Berger, VW, Yu, Z (2017). The asymptotic maximal procedure for subject randomization in clinical trials. *Statistical Methods in Medical Research*, in press.

5

Restricted Randomization: Pros and Cautions

Jonathan Chipman

CONTENTS

5.1 Introduction 51
5.2 Historical Context of Randomization 52
5.3 Restricted Randomization to Control Biases 53
5.4 Covariate-Adjusted Restricted Randomization to Control Biases
and Increase Precision Estimating the Experimental Condition's
Effect 55
5.5 Cautions (Considerations) of Covariate-Adjusted Randomization
Approaches 56
5.6 Case Studies and Conclusions 57
References 59

5.1 Introduction

This chapter looks back to the foundation of randomization and ties together modern-day randomization approaches and considerations. Randomization removes, on average, the effect of any bias upon the experiment's conclusions. Yet, in a single randomization, nontrivial imbalances in factors related to the outcome (i.e. confounders) can be common, and rare randomizations can fail dramatically. Restrictions to randomization help diminish the risk of rare, bad randomizations and can increase the efficiency of the experiment to detect the effect of an experimental condition, if one truly exists. The emphasis of this chapter is to understand the context of randomization, bring awareness of pros and cautions of various randomization schemes, and encourage covariate-adjusted randomization in experiments where experimental units are randomized sequentially. There is a wealth of additional literature on randomized trials where experimental units are all known before randomization and on covariate-adjusted response-adaptive randomization, though these topics are beyond the scope of this chapter.

5.2 Historical Context of Randomization

Charles Peirce[1] and RA Fisher[2] are often considered the founders of randomization in controlled experiments. Peirce used randomization to remove his own subjective biases from influencing the outcome, adding objective validity to his studies. Fisher also advocated for randomization to achieve unbiased estimates and further used randomization for significance testing—a statistical approach to describe how improbable the experiment's outcome, or one more extreme, would be if the experimental condition were truly ineffective.[2] To Peirce's objective validity, Fisher added the concept of statistical significance. In turn, randomization became a tool for assessing an estimator's uncertainty due to sampling variation (standard error).

Though randomization for controlled experiments faced initial criticism, that it was unnecessary and inconvenient,[3] it has since become the gold standard. And, quite impressively, Peirce and Fisher used randomization approaches that are still among the most commonly used schemes today. Indeed, there are multiple approaches for obtaining a randomization design (the random ordering of experimental condition assignments). For an experiment with two experimental conditions, the most simplistic approach is to flip a coin for each experimental unit (such as for each study participant). This may result in rare but very bad designs, such as only one experimental condition being randomized to each experimental unit. Such an experiment wouldn't even make a comparison! In 1992, C. Radhakrishna Rao reflected on Fisher's advances in randomization and opened a discussion applicable to modern-day randomization:

> However, a few points remain to be resolved in the practice and theory of randomization. What should one do if a design arrived at by random choice exhibits systematic features? Should one reject this and make another random choice? Any design of experiment must specify the set of designs from which one may be chosen at random. The only condition the set has to satisfy is that the act of randomization provides unbiased estimates of treatment comparisons and valid estimates of error. If there is more than one set with these properties, what further criteria should be used in choosing an appropriate set?[4]

Rao raises two basic questions. First, can an experimenter selectively choose between multiple randomized designs? Morgan and Rubin later addressed this question by providing the statistical justification and considerations for rerandomization in experiments where all experimental units are known before randomization.[5] Alternatively, various randomization restrictions are often imposed to diminish the risk of obtaining a randomized design with systematic features. Rao's second question, asking what properties should guide an experimenter in choosing between multiple valid approaches to randomization, becomes important.

Randomization schemes are often compared for its properties to control biases, ability to increase the precision of the estimated experimental condition's effect, ability to achieve a desired allocation ratio of experimental conditions, and available analytic methods. Though biases can come in many ways, randomization schemes often protect against assigning earlier experimental units to one experimental condition and vice versa (chronological bias), selecting experimental units based on correctly guessing the next randomized experimental condition (subversion bias), and masking the experimental condition's effect due to imbalanced confounders (confounding bias). Altogether, these properties provide a basis for comparing some common modern approaches to randomization.

5.3 Restricted Randomization to Control Biases

Peirce provided the first example of what today we describe as Block Randomization. In an experiment to determine whether individuals could recognize subtle differences between weight, Peirce used a deck of playing cards to either add or remove weight from a scale counterbalanced by his blinded colleague.[1] By the end of the experiment, an equal number of replications were administered for the different experimental conditions. This practice places a restriction upon randomization by eliminating any design with unequal replicates of experimental conditions. In Block Randomization, blocks describe the number of experimental units for which an equal allocation of experimental conditions is forced. Using small block sizes, the experiment may be balanced in experimental condition replicates at regular intervals of the study.

As a strength, Block Randomization helps control against the chronological bias. Yet, in so doing, Block Randomization may open the door to subversion bias, which in turn may inflate Type I error and confounding bias.[6] Subversion bias may occur in clinical studies where the person enrolling participants may anticipate the upcoming treatment assignment. They might allow their subjective bias to not enroll a participant until later when there is a greater chance for the other experimental condition to be assigned. Permuted Block Randomization uses random ordering of different sized blocks, which makes subversion more challenging. Still, the risk of subversion is real in any study where either the receiver or administrator of the experimental condition is unblinded (able to determine the experimental condition). In 2011, Zhao and Weng observed from clinicaltrials.gov that slightly more than half of 5097 currently recruiting Phase III trials were open label or singly blind.[7]

Efron's Biased Coin[8] and LJ Wei's Urn Randomization[9,10] both have the same goal of achieving equal, or nearly equal, allocation of experimental conditions. Neither restricts the possible set of randomization designs, but the chance of some designs occurring are so negligible that these schemes are considered restricted randomization. They are also dynamic randomization schemes because they allow the probability of assigning an experimental condition to change throughout the study. Biased Coin Randomization assigns an experimental condition with some favorable probability, p, to the experimental condition with fewer assigned experimental units. Though p must be determined before the study begins, Urn Randomization provides an algorithm for assigning p based on the imbalance of assigned experimental conditions. An experimenter has a figurative urn (basket) of balls, each colored to represent an experimental condition. When the first ball is drawn, there are an equal number of each colored ball. The drawn ball indicates the experimental condition assignment; the ball is placed back into the urn; and additional β balls corresponding to the other experimental condition's color are added to the urn. As the study continues, the biased coin for assigning experimental conditions becomes less biased.

Whereas the Biased Coin and Urn Randomization prevent against subversion and aim to achieve nearly equal experimental condition assignment, a critique remains that a chronological bias may still occur.[7] Maximum Tolerable Indifference (MTI) designs set an upper limit of the imbalance of assigned experimental conditions at any single point of an experiment.[11] Permuted Block Randomization falls within the MTI umbrella as well as Maximal Procedure and Balanced Urn Design. Maximal Procedure considers the set of randomizations that would have an imbalance no greater than a pre-specified amount at any given point in time.[11] And, the Balanced Urn Design makes use of a secondary urn to help maintain allocation balance of experimental conditions. Compared to Permuted Blocks, Maximal Procedure and Balanced Urn Design randomization schemes have substantially fewer deterministic experimental conditions and substantially less risk of guessing an upcoming experimental condition's assignment.[7]

Permuted Block (when without subversion), Biased Coin, and Urn Randomization will balance out confounding factors, on average, yet any single design may result in experimental condition groups that are different in some confounder that could compromise the conclusion. Consider, for example, a clinical trial design that by chance randomizes the healthiest participants to the novel treatment. At the end of the study, the experimenter would have a hard time concluding whether the health status or treatment brought about a more favorable outcome. By adjusting randomization to potential confounders, the experimenter reduces the risk of confounded conclusions and can increase the precision of the estimated outcome—a key property for comparing randomization schemes. The risk of confounder imbalances is greater in smaller studies; however, Greevy showed that regardless of the experiment's size, adjusting a linear model to more well-balanced covariates has a nonnegative gain in precision for estimating an experimental condition's effect using a covariate-adjusted linear model.[12]

5.4 Covariate-Adjusted Restricted Randomization to Control Biases and Increase Precision Estimating the Experimental Condition's Effect

Fisher conducted experiments using factorial designs to increase the precision of experimental outcomes and to further study how key factors (covariates) may jointly modify the outcome.[2] To carry out a factorial design, an experimenter denotes all possible levels of each key confounding covariate and randomizes experimental units within each combination covariate levels. Stratified Block Randomization works in this manner. An experimenter that would want to control for three levels of age (young, middle, and elderly), gender (male and female), and race (white, black, and other) would randomize patients within $3 \times 2 \times 3 = 18$ strata.

Stratified Block Randomization is a mainstay in randomized trials. It is, however, limited in controlling for only a few categorical covariates. As the number of categorical covariate levels increases, so too increases the number of strata to randomize participants. The largest number of strata to still guarantee equal allocation of two experimental conditions within-strata is half the enrollment size. Twin studies, for example, randomize each member of the twin set (the strata) to two different experimental conditions. As a consequence, Stratified Block Randomization quickly becomes limited in the amount of categorical covariates used to restrict randomization. Often experiments implement Permuted Block Randomization within strata to achieve an equal allocation of experimental units. Still, even with few strata, strata may remain imbalanced in the experimental condition assignment and induce a confounding covariate in the overall study.

Matched Randomization allows for adjusting randomization to any number of continuous and categorical covariates.[12] The randomization scheme takes root in observational studies in which one might use matching to compare a set of cases and controls balanced on all important covariates. In studies where all experimental units are known before randomization, the experimenter uses a distance measure, such as Mahalanobis Distance, to identify the set of matches that are overall most similar to one another in terms of the continuous and categorical covariates. Then, the experimental condition is randomized to either experimental unit of the match. In this way, Matched Randomization is similar to a twin study. Matched Randomization has further been extended for when experimental units enter sequentially into the experiment.[13] Thus far, Matched Randomization has been developed for experiments with two equally allocated experimental conditions; however, this is an area of ongoing research.

Though also limited to adjusting for categorical covariates, Urn Randomization can adjust for more covariates than Stratified Block Randomization because it balances covariates marginally.[10] Each level of a categorical covariate is assigned its own urn. In the example of controlling for age, gender, and race, there would be 3 + 2 + 3 = 8 urns. Then, when an experimental unit needs to be randomized, the experimenter draws a ball from the urn corresponding to the experimental unit's least well-balanced covariate level. If a young, male, and white study participant needed randomization, a ball would be drawn from the young urn if the allocation balance of experimental conditions were worse among young participants than among male or white participants. The experimenter would then return the drawn ball and add β balls of the opposite experimental condition to the randomized experimental unit's covariate levels (in this example to young, male, and white urns).

Similar in its ability to marginally adjust for multiple covariates, Taves proposed Minimization,[14] which deterministically assigns the experimental condition that would yield the least amount of covariate imbalance. When an experimental unit requires an experimental condition, the experimenter first calculates a measure of covariate deviance that would result from it receiving either experimental condition. An experimental condition is then assigned to whichever experimental condition would lead to less deviant imbalance. Begg and Iglewicz as well as Atkinson proposed assigning experimental conditions that would estimate the experimental condition's effect with least variability using a covariate-adjusted multivariate regression model.[15,16] Pure Minimization deterministically allocates the experimental condition that minimizes the deviance measure yet has consequently been critiqued as leaving the experiment vulnerable to subversion bias.[17,18] However, Simon and Pocock extended Minimization to allow randomization favoring the experimental condition that would result in minimizing deviant imbalance.[19]

5.5 Cautions (Considerations) of Covariate-Adjusted Randomization Approaches

Fisher argued for randomization because it provides a valid significance test when all key covariates are balanced between experimental conditions (i.e. in the absence of confounders). When confounders exist, the analysis must adjust for these covariates such as with a multivariate linear regression model. Even in the absence of confounders, many experimenters will analyze their study with a multivariate regression model to increase the estimated precision of the experimental condition's effect; however, they must assume the model assumptions are met. Often an unadjusted randomization test is provided to show robustness of multivariate regression modeling assumptions, and concern may be raised if conclusions from the adjusted and unadjusted analysis do not agree. Under Stratified and Matched Randomization, an experimenter may carry out both an unadjusted and multivariate linear model analysis.[13] Yet, because experimental conditions under Minimization,

Biased Coin, and Urn Randomization depend in a complex manner upon past experimental condition allocations, an unadjusted analysis will not yield accurate standard errors for comparison. These randomization schemes require the analysis to adjust for covariates used in randomization. In large studies, however, Urn Randomization becomes increasingly more like an unrestricted randomization design which may justify an unadjusted randomization test.[10]

Stratified Block Randomization with Permuted Blocks can control for chronological bias though this may be jeopardized if the number of strata are too large. However, subversion remains a concern in studies without blinding and when experimenters know the current block size. On the other hand, the opposite is true for Matched Randomization, Urn Randomization, and Minimization Randomization which do not fall under the umbrella of MTI. There remains a chance for chronological bias. Yet, adjusting for a factor that represents time of enrollments may help mitigate the chronological bias. In unblinded studies and in using randomization, these later randomization schemes prevent against concerns of subversion by not imposing deterministic allocation of experimental conditions.

As to the efficiency of covariate-adjusted randomization approaches, Kapelner and Krieger found in simulations that Matched Randomization and Minimization yield substantially greater power to detect an experimental condition's effect than Stratified Block Randomization when using a covariate-adjusted linear model for analysis.[13] Whereas Stratified Block and Urn Randomization require categorical covariates for adjusting randomization, Matched Randomization and Minimization do not carry this constraint. Categorizing continuous potential confounders, such as age, always comes with a loss of important information.

In designing a covariate-adjusted randomized study, an experimenter must be mindful of the number of covariates used to adjust randomization. Even in Urn Randomization, Minimization, and Matched Randomization schemes, balancing on many additional weak potential confounders may take away from the benefit of balancing on a few strong potential confounders. In using Matched Randomization, an experimenter may use a weighted version of Mahalanobis Distance to prioritize the importance of balancing confounders,[20] and careful weighting of the deviance measure can prioritize balancing important confounders using Minimization.

5.6 Case Studies and Conclusions

Echoing Fisher's principles of a well-designed study, the Federal Drug Administration recommends designing studies that minimize bias and maximize efficiency.[21] Whenever possible, blinding should be implemented.

Furthermore, restricted randomization, with particular emphasis on covariate-adjusted randomization, can help reduce confounding biases and may increase the efficiency of a study. Urn Randomization allows for adjusting for more categorical confounders than Stratified Block Randomization, and Matched Randomization and Minimization further allow for adjusting unlimited continuous confounders. Priority may be given to more important confounders by appropriate weighting in Matched Randomization and Minimization. In unblinded studies, consideration must be given to the risk and prevention of subversion bias.

Reported in 2004, the ACE Inhibitor After Anthracycline study[22] could have benefited from Matched Randomization, Minimization, or covariate-adjusted Urn Randomization. The study aimed to estimate the effect to which enalapril, an angiotensin-converting enzyme (ACE), prevented cardiac function decline among individuals exposed to anthracyclines as children. In the reported Table 1, 12 covariates were relevant to the outcome, of which 7 were categorical. Using Stratified Block Randomization, and being limited in the number of categorical covariates they could stratify upon, they adjusted for dichotomized versions of age at treatment, total cumulative anthracycline dosage, and time from diagnosis for a total of 8 strata. Still, the randomization left 9 of 10 black participants randomized to placebo among the 135 randomized participants. Even a linear model would be limited in adjusting for race because the participants assigned to enalapril were almost all of one race. After the completion, the trial faces the question of how much of the enalapril effect was due to race being imbalanced among treatment groups. While covariate-adjusted Urn Randomization would have helped in this situation, Matched Randomization and Minimization could have further avoided categorizing continuous factors such as age, dosage, and time from diagnosis.

Large experiments are not immune to confounding bias. The Breast Cancer Erythropoietin Trial, reported in 2003, randomized 939 metastatic breast cancer patients from 139 international sites to test the hypothesis that the drug erythropoietin led to improved 1-year survival.[23] Surprisingly, the study terminated because those on placebo had a 76% 1-year survival rate compared to a 70% survival rate among those treated. In reporting the findings, the Principal Investigator communicated that "the results of this trial must be interpreted with caution in light of the potential for an imbalance of risk factors between treatment groups."[23] The study team did not collect important prognostic factors on all patients, so an adequately adjusting linear model could not be fit. Covariate-adjusted randomization would have helped the study team not only balance on important prognostic factors but also adequately plan ahead. The Principal Investigator further lamented, "It is extremely unfortunate that the problems in design, conduct, and post-trial analysis have complicated the interpretation of this study."[23] In 2009, the European Organisation for Research and Treatment of Cancer Working Group reviewed multiple clinical trials and meta-analysis on the benefit of erythropoiesis-stimulating agents and recommended their favorable use.[24] Regardless of the true effect, a significant amount of patients and resources were dedicated to the study for ultimately a confounded conclusion.

Given the pros of covariate-adjusted randomization, with careful consideration to implementation and subsequent analysis, why then is it not more commonly utilized? Similar as in Fisher's day, a common argument is that in large studies it is inconvenient and unnecessary. Certainly, there is a case for using covariate-adjusted randomization in smaller studies where the risk for confounding and need for added precision is greatest. Still, even in large studies, there is always a benefit of covariate-adjusted randomization. Matched Randomization, for example, has been shown to have a non-negative increase to the relative efficiency by as much as 7% when using an adjusted linear model for analysis.[12] Covariate-adjusted randomization can be fairly easily implemented in software databases to make randomization assignments immediate once covariates have been recorded,[25] and considering the cost of large trials, having added protection against covariate imbalance and increased precision seems well worth the added effort.

References

1. Peirce CS, Jastrow J. On small differences in sensation [Internet]. 1884. Available from: http://psychclassics.yorku.ca/Peirce/small-diffs.htm.
2. Fisher RA. *The Design of Experiments.* 1935. Oliver and Boyd, Edinburgh.
3. Hall NS. R. A. Fisher and his advocacy of randomization. *Journal of the History of Biology* 2007;40(2):295–325.
4. Rao CR. R. A. Fisher: The Founder of Modern Statistics. *Statistical Science* 1992; 7(1):34–48.
5. Morgan KL, Rubin DB. Rerandomization to improve covariate balance in experiments. *The Annals of Statistics* 2012.
6. Berger VW, Weinstein S. Ensuring the comparability of comparison groups: Is randomization enough? *Controlled Clinical Trials* 2004;25(5):515–24.
7. Zhao W, Weng Y. Block urn design—A new randomization algorithm for sequential trials with two or more treatments and balanced or unbalanced allocation. *Contemporary Clinical Trials* 2011;32(6):953–61.
8. Efron B. Forcing a sequential experiment to be balanced. *Biometrika* 1971;58(3):403.
9. Wei L-J. A class of designs for sequential clinical trials. *Journal of the American Statistical Association* 1977;72(358):382.
10. Wei LJ, Lachin JM. Properties of the urn randomization in clinical trials. *Controlled Clinical Trials* 1988;9(4):345–64.
11. Berger VW, Ivanova A, Deloria Knoll M. Minimizing predictability while retaining balance through the use of less restrictive randomization procedures. *Statistics in Medicine* 2003;22(19):3017–28.
12. Greevy R, Lu B, Silber JH, Rosenbaum P. Optimal multivariate matching before randomization. *Biostatistics* 2004;5(2):263–75.

13. Kapelner A, Krieger A. Matching on-the-fly: Sequential allocation with higher power and efficiency. *Biometrics* 2014;70(2):378–88.

14. Taves DR. Minimization: A new method of assigning patients to treatment and control groups. *Clinical Pharmacology & Therapeutics* 1974;15(5):443–53.

15. Begg CB, Iglewicz B. A treatment allocation procedure for sequential clinical trials. *Biometrics* 1980;36(1):81.

16. Atkinson AC. Optimum biased coin designs for sequential clinical trials with prognostic factors. *Biometrika* 1982;69(1):61.

17. Taves DR. The use of minimization in clinical trials. *Contemporary Clinical Trials* 2010;31(2):180–4.

18. Berger VW. Minimization, by its nature, precludes allocation concealment, and invites selection bias. *Contemporary Clinical Trials* 2010.

19. Pocock SJ, Simon R. Sequential treatment assignment with balancing for prognostic factors in the controlled clinical trial. *Biometrics* 1975;31(1):103–15.

20. Greevy RA Jr., Grijalva CG, Roumie CL et al. Reweighted Mahalanobis distance matching for cluster-randomized trials with missing data. *Pharmacoepidemiology Drug Safety* 2012;21:148–54.

21. Guidance for industry: Adaptive design clinical trials for drugs and biologics [excerpts]. *Biotechnology Law Report* 2010;29(2):197–215.

22. Silber JH, Cnaan A, Clark BJ et al. Enalapril to prevent cardiac function decline in long-term survivors of pediatric cancer exposed to anthracyclines. *Journal of Clinical Oncology* 2016;22(5):820–8.

23. Leyland-Jones B. Breast cancer trial with erythropoietin terminated unexpectedly. *The Lancet Oncology* 2003.

24. Aapro M, Spivak JL. Update on erythropoiesis-stimulating agents and clinical trials in oncology. *The Oncologist* 2009;14 Suppl 1(Supplement 1):6–15.

25. Lange N, MacIntyre J. A computerized patient registration and treatment randomization system for multi-institutional clinical trials. *Controlled Clinical Trials* 1985;6(1):38–50.

6

Evolution of Restricted Randomization with Maximum Tolerated Imbalance

Wenle Zhao

CONTENTS

6.1   Introduction 61
6.2   The Conditional Allocation Probability 62
    6.2.1   The Permuted Block Design 63
    6.2.2   The Block Urn Design 64
    6.2.3   The Big Stick Design 64
    6.2.4   The Biased Coin Design with Imbalance Tolerance 65
    6.2.5   The Maximal Procedure 65
    6.2.6   The Asymptotic Maximal Procedure 67
    6.2.7   Comparison of Conditional Allocation Probabilities 72
6.3   The Treatment Allocation Predictability 76
6.4   Implementation of MTI Procedures 79
6.5   Summary 79
References 80

6.1 Introduction

The term *maximum tolerated imbalance* (MTI) was first used by Vance Berger et al. in 2003 when the maximal procedure was proposed as a contrast to the permuted block design.[1] The maximal procedure replaces the enforced balance at the end of each block with the enforced balance at the end of the allocation sequence, aiming to reduce the treatment assignment predictability by increasing the number of feasible allocation sequences.[1,2]

To avoid the unwanted imbalances resulting from the complete randomization, various restricted randomization procedures have been proposed aiming to have equal number of subjects assigned to each treatment arm.[3] A restricted randomization can be categorized as an MTI procedure if a pre-specified treatment imbalance limit is enforced by deterministic assignments.[4] Therefore, the big stick design proposed by Soares and Wu,[5] the biased coin design with imbalance tolerance proposed by Chen,[6] the maximal procedure proposed by Berger et al.,[1] the block urn design proposed by

Zhao and Weng,[7] and the asymptotic maximal procedure proposed by Zhao et al.[8] are all MTI procedures. On the other hand, Efron's biased coin design,[9] Wei's urn design,[10] Smith's generalized biased coin design,[11] Chen's Ehrenfest urn design,[12] and Antognini's symmetric and asymmetric Ehrenfest urn designs[13] are not MTI procedures.

Initially proposed by Hill,[14] the permuted block design is by far the most commonly used randomization method in clinical trial practice.[15] It uses deterministic assignments not only when the treatment imbalance reaches its maximum tolerated level defined by the block size but also when the allocation sequence approaches the end of each block.[16] We classify the permuted block design as a special MTI procedure and include it as the benchmark for the evaluation of the statistical properties of other MTI procedures.

In this chapter, the mechanisms of the aforementioned six MTI procedures will be illustrated based on the conditional allocation probability. Their performance will be compared with each other based on the treatment assignment predictability. Finally, a generic implementation strategy of MTI procedures in sequential clinical trials is provided.

6.2 The Conditional Allocation Probability

The primary goal of restricted randomization is to control the treatment distribution imbalance. It is a natural strategy to alter the treatment allocation probability according to the observed treatment distribution. A restricted randomization procedure can be classified as a Markov process if the treatment assignment for the next subject depends solely on the current treatment distribution, not on the treatment assignment sequences that preceded it.[8] Consider a two-arm trial with equal allocation, let $n_{i,A}$ and $n_{i,B}$ be the number of subjects previously assigned to arms A and B, respectively, among the i subjects being randomized into the study. Let $p_{i,A}$ and $p_{i,B}$ be the conditional allocation probability for assigning subject i to arms A and B, respectively, and R_i be the random number with uniform distribution on $(0,1)$ for the randomization of subject i. The conditional allocation probability for a restricted randomization takes a general format as below:

$$p_{i,j} = F(n_{i-1,A}, n_{i-1,B}, \mathbf{C}) \quad (i = 1, 2, \cdots; j = A, B), \tag{6.1}$$

where $\mathbf{C}$ is the vector of constants used by the randomization algorithm, such as the MTI, the block size for the permuted block design and the block urn design, or the biased coin probability used by Chen's biased coin design with imbalance tolerance. If the difference between the sizes of the two arms, rather than the sizes themselves, is the concern, the conditional allocation probability can be simplified to

$$p_{i,j} = F(d_{i-1}, \mathbf{C}) \quad (d_{i-1} = n_{i-1,A} - n_{i-1,B}; \, i = 1, 2, \cdots; \, j = A, B). \tag{6.2}$$

Zhao et al. compared treatment imbalance and allocation predictability for 14 commonly used randomization designs, including a few MTI procedures, based on the conditional allocation probability under a wide range of trial setting scenarios.[17]

6.2.1 The Permuted Block Design

The conditional allocation probability for the *permuted block design* (PBD) can be obtained by an urn model.[18] Consider an m-arm trial with a target allocation of $r_1 : r_2 : \cdots : r_m$, where r_j ($j = 1, 2, \cdots m$) are positive integers without common divisors greater than 1. Let $r^* = \sum_{j=1}^{m} r_j$ be the sum of allocation ratio elements, and $b = \lambda r^*$ be the block size, a positive multiple of r^*. The urn starts from b balls, with λr_j balls for arm $j = 1, 2, \cdots m$. The randomization procedure is conducted based on a random draw of a ball from the urn without replacement, and the subject is assigned to the treatment arm corresponding to the selected ball. When the urn is empty, all balls are returned to the urn to start a new block. Let n_j ($j = 1, 2, \cdots m$) be the number of subjects assigned to arm j among the first i subjects. The probability for assigning subject i to arm j is

$$p_{i,j}(\text{PBD}) = \frac{\lambda r_j + k_{i-1}\lambda r_j - n_{i-1,j}}{b + bk_{i-1} - (i-1)} \quad (i = 1, 2, \cdots; j = 1, 2, \cdots m). \tag{6.3}$$

Here, $k_{i-1} = \text{int}[(i-1)/b]$ is the number of completed blocks in the $(i-1)$ assignments, and $\text{int}(x)$ returns the largest integer less than or equal to x. For two-arm equal allocation trials, Equation 6.3 is reduced to

$$p_{i,A}(\text{PBD} \mid \text{two-arm 1:1 allocation}) = \frac{\text{int}[(i-1)/b+1] \times b/2 - n_{i-1,A}}{\text{int}[(i-1)/b+1] \times b - (i-1)} \quad (i = 1, 2, \cdots). \tag{6.4}$$

Both Equations 6.3 and 6.4 indicate that the conditional allocation probability of the permuted block design depends on the current treatment distribution, not solely on the current treatment imbalance. For example, based on Equation 6.4, for a two-arm equal allocation trial with block size $b = 6$, when $n_{10,A} = 6$ and $n_{10,B} = 4$, there is $p_{11,A} = 0$; and when $n_{14,A} = 8$ and $n_{14,B} = 6$, there is $p_{15,A} = 1/4$, although the treatment imbalances in both scenarios equal to 2.

6.2.2 The Block Urn Design

The *block urn design* (BUD) was proposed by Zhao and Weng in 2011 as a better alternative to the permuted block design.[7] It uses the same urn model as the permuted block design with a different ball return policy. Instead of waiting until the end of a block to return all b balls, a minimum set of balanced balls is returned as soon as they become available. This minimum set has r^* balls, with r_j ($j = 1,2,...m$) balls for arm j. Let $k^* = \min_{1 \leq j \leq m}[\text{int}(n_{i-1,j}/r_j)]$ be the number of balanced sets of assignments among the $(i-1)$ previously randomized subjects. The conditional allocation probability is

$$p_{i,j}(\text{BUD}) = \frac{\lambda r_j + k_{i-1}^* r_j - n_{i-1,j}}{b + k_{i-1}^* r^* - (i-1)} \quad (i = 1, 2, \cdots; j = 1, 2, \cdots m). \tag{6.5}$$

For two-arm equal allocation trials, there is $k_{i-1}^* = \min(n_{i-1,A}, n_{i-1,B})$, and Equation 6.5 is reduced to

$$p_{i,A}(\text{BUD} | \text{two-arm } 1:1) = \frac{b/2 + k_{i-1}^* - n_{i-1,A}}{b + 2k_{i-1}^* - (i-1)} \quad (i = 1, 2, \cdots). \tag{6.6}$$

With $(i-1) - 2k_{i-1}^* = n_{i-1,A} + n_{i-1,B} - 2\min(n_{i-1,A}, n_{i-1,B}) = |n_{i-1,A}, n_{i-1,B}| = |d_{i-1}|$, Equation 6.6 can be further simplified to

$$p_{i,A}(\text{BUD} | \text{two-arm } 1:1) = \frac{1}{2} - \frac{d_{i-1}}{2(b - |d_{i-1}|)} \quad (i = 1, 2, \cdots). \tag{6.7}$$

This result indicates that, for two-arm equal allocation trials, the conditional allocation probability of the block urn design depends on the current treatment imbalance only. It is not difficult to prove that the block urn design and the permuted block design are identical when the minimally possible block size is used. In that case, there are $\lambda = 1$ and $b = r_1 + r_2$.

6.2.3 The Big Stick Design

Proposed by Soares and Wu in 1983 for two-arm balanced allocation trials, the *big stick design* (BSD) uses complete random assignments until the MTI is reached.[5] It has a conditional allocation probability as below:

$$p_{i,A}(\text{BSD}) = \begin{cases} 0 & \text{when } d_{i-1} = \text{MTI} \\ 1/2 & \text{when } |d_{i-1}| < \text{MTI} \quad (d_{i-1} = n_{i-1,A} - n_{i-1,B}) \\ 1 & \text{when } d_{i-1} = -\text{MTI} \end{cases} \tag{6.8}$$

6.2.4 The Biased Coin Design with Imbalance Tolerance

Chen proposed the *biased coin design with imbalance tolerance* (BCDWIT) in 1999, which incorporated an imbalance limit, that is, the MTI, into Efron's original biased coin design.[6,9] It could also be considered as a modified version of the big stick design by introducing a fixed biased coin probability before reaching the MTI. The conditional allocation probability of this randomization design depends fully on the current treatment imbalance:

$$p_{i,A}(\text{BCDWIT}) = \begin{cases} 0 & \text{when } d_{i-1} = \text{MTI} \\ 1 - p_{bc} & \text{when } 0 < d_{i-1} < \text{MTI} \\ 1/2 & \text{when } d_{i-1} = 0 \\ p_{bc} & \text{when } 0 > d_{i-1} > -\text{MTI} \\ 1 & \text{when } d_{i-1} = -\text{MTI} \end{cases} \qquad (d_{i-1} = n_{i-1,A} - n_{i-1,B})$$

$$(6.9)$$

Here, $p_{bc} > 1/2$ is a biased coin probability in favor of assigning the subject to the smaller arm. Currently, applications of the big stick design and the biased coin design with imbalance tolerance are limited to two-arm equal allocation trials only. Generalizations of the big stick design and the biased coin design with imbalance tolerance to more than two arms and/or unequal allocation trials are not available, primarily due to the lack of corresponding definitions of MTI.

6.2.5 The Maximal Procedure

Vance et al. proposed the *maximal procedure* (MP) in 2003, assigning an equal probability to all feasible allocation sequences under four constraints:[1]

1. The target allocation ratio, such as (r_1, r_2)
2. The sequence length N
3. The maximal tolerated imbalance $\text{MTI} = \max_{1 \leq i \leq N} (|d_i|)$
4. The perfect terminal balance $d_N = 0$

The equal allocation sequence probability is a desirable statistical property for a randomization procedure, especially when randomization tests of the trial results are considered. The maximal procedure can be applied to equal and unequal allocation trials. Theoretically speaking, it could also be applied to trials with more than two arms. In 2008, Salama et al. published an algorithm for the construction of the maximal procedure allocation sequences using a graphical strategy.[2] As depicted in Figure 6.1, $s(n_A, n_B)$ represents the number of feasible allocation sequences from node (n_A, n_B) to the terminal

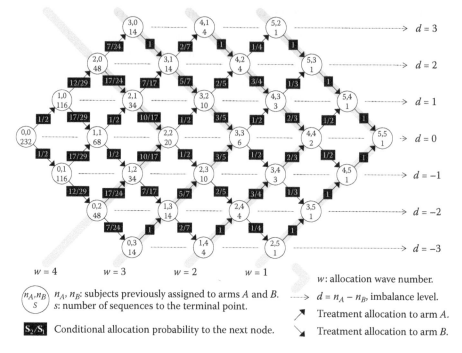

FIGURE 6.1
Maximal procedure feasible allocation sequences and conditional allocation probabilities (two-arm equal allocation, MTI = 3, sequence length $N = 10$).

node. All feasible sequences from node (n_A, n_B) must pass through one of the two possible next nodes $(n_A + 1, n_B)$ or $(n_A, n_B + 1)$.

This reasoning strategy leads to the following procedure (Equations 6.10 through 6.13) for the calculation of the maximal procedure conditional allocation probability for two-arm equal allocation trials:

$$s(\text{terminal node}) = 1 \tag{6.10}$$

$$s(n_A, n_B) = 0 \text{ when } |n_A - n_B| > \text{MTI} \tag{6.11}$$

$$s(n_A, n_B) = s(n_A + 1, n_B) + s(n_A, n_B + 1) \tag{6.12}$$

$$p_A(\text{MP} \mid n_A, n_B) = \frac{s(n_A + 1, n_B)}{s(n_A, n_B)}. \tag{6.13}$$

Equation 6.13 indicates that the conditional allocation probability from one node to the next node is proportional to the number of feasible allocation sequences passing through that next node to the terminal node. This logic ensures that all feasible allocation sequences from the start node to the terminal node will have equal probabilities to be selected by this randomization algorithm. Condition 6.11 can be modified for two-arm unequal allocations.[1,2,8] For example, for a two-arm trial with an allocation of 3:2, Condition 6.11 can be modified to $s(n_A, n_B) = 0$ when $|2n_A - 3n_B| > $ MTI. Table 6.1 illustrates the generation of a sample maximal procedure allocation sequence.

A short maximal procedure randomization sequence for a two-arm equal allocation trial, like the one in Table 6.1, can be manually calculated based on formulas (Equations 6.10 through 6.13). Longer sequences for multi-arm and/ or unequal allocation trials may require the help of a computer program, or at least a spreadsheet. The maximal procedure does not have the Markovian property, and therefore, there is no explicit conditional allocation probability formula, in the form of Equation 6.1 or 6.2, making the implementation of the maximal procedure difficult.

6.2.6 The Asymptotic Maximal Procedure

In 2016, Zhao et al. proposed the *asymptotic maximal procedure* (AMP),[8] by lifting the allocation sequence length constraint N and the terminal balance constraint in the maximal procedure $d_N = 0$. In sequential clinical trial practice, restricted randomization is often stratified by baseline covariate categories. The length of an allocation sequence N, that is, the number of subjects in the corresponding baseline covariate stratum, remains unknown before the completion of the study. Therefore, both the pre-specified allocation sequence length N and the terminal balance requirement $d_N = 0$ are not relevant in practice. Furthermore, the lack of the Markovian property makes the implementation of the maximal procedure more difficult than other MTI procedures. For these reasons, the asymptotic maximal procedure was developed under the conditions of the target allocation ratio and the MTI, while keeping the probability equal for all feasible allocation sequences.

As depicted in Figure 6.1, let N denote the sequence length and n_A and n_B represent the number of subjects assigned to arms A and B, respectively, before node (n_A, n_B). Let $d = n_A - n_B$ be the treatment imbalance level. Define $w \geq 0$ as the *allocation wave* number, which equals the total number of nodes with the same imbalance level as the current node before the terminal point. For example, in Figure 6.1, node (3,2) has imbalance level $d = 1$. Between node (3,2) and terminal point (5,5), there are two nodes, that is, (4,3) and (5,4), with the same imbalance level of 1. Therefore, node (3,2) has $w = 2$. The group of all allocation nodes with the same w value is considered as an allocation wave. A complete allocation wave always consists of (2MTI + 1) nodes, one for each tolerated imbalance level d. For example, wave 1 consists of nodes (1,4), (2,4), (3,4), (4,4), (4,3), (4,2), and (4,1). For two-arm balanced allocation

TABLE 6.1

Sample Maximal Procedure Allocation Sequence Generation (Two-Arm Equal Allocation Trial, Sequence Length $N = 10$, MTI = 3)

Sequence Number	Current Node (n_A, n_B)	Next Node $(n_A + 1, n_B)$	Next Node $(n_A, n_B + 1)$	Sequences through $(n_A + 1, n_B)$	Sequences through $(n_A, n_B + 1)$	Conditional Allocation Probability $P(T = A)$	Random Number	Treatment Assignment T
1	(0,0)	(1,0)	(0,1)	116	116	0.5	0.7109	B
2	(0,1)	(1,1)	(0,2)	68	48	0.586207	0.9014	B
3	(0,2)	(1,2)	(0,3)	34	14	0.708333	0.8714	B
4	(0,3)	(1,3)	–	14	0	1	0.1541	A[a]
5	(1,3)	(2,3)	(1,4)	10	6	0.714286	0.3398	A
6	(2,3)	(3,3)	(2,4)	6	4	0.6	0.3617	A
7	(3,3)	(4,3)	(3,4)	3	3	0.5	0.4293	A
8	(4,3)	(5,3)	(4,4)	1	2	0.333333	0.2771	A
9	(5,3)	–	(5,4)	0	1	0	0.2176	B[a]
10	(5,4)	–	(5,5)	0	1	0	0.1850	B[a]

Note: (n_A, n_B), allocation node with n_A and n_B subjects previously enrolled in arms A and B, respectively.

[a] Deterministic assignment.

trials, the allocation wave number w equals $\min(N/2 - n_A, N/2 - n_B)$. The combination of the allocation wave number w and the imbalance level d uniquely identifies an allocation node. For this reason, an allocation node can be interchangeably represented by (n_A, n_B) or node$_{w,d}$. For example, node $(1,2)$ is equivalent to node$_{3,-1}$. Node$_{0,0}$ represents the balanced terminal node.

Denote $s_{w,d}$ as the number of feasible sequences from node$_{w,d}$ to the terminal node$_{0,0}$. Let $p_{w,d,A}$ and $p_{w,d,B}$ be the conditional allocation probabilities at node$_{w,d}$ for assigning the current subject to treatments A and B, respectively. For two-arm balanced trials, the number of feasible allocation sequences from node$_{w,d}$ to the terminal node$_{0,0}$ can be obtained from a recursive function of parameters w and d:

$$
s_{w,d} = \begin{cases}
1 & w = 0; d \neq 0 \\
s_{w,1-\text{MTI}} & w > 0; d = -\text{MTI} \\
s_{w,d+1} + s_{w-1,d-1} & w > 0; -\text{MTI} < d < 0 \\
s_{w-1,-1} + s_{w-1,1} & w > 0; d = 0 \\
s_{w,d-1} + s_{w-1,d+1} & w > 0; 0 < d < \text{MTI} \\
s_{w,\text{MTI}-1} & w > 0; d = \text{MTI}
\end{cases}
\tag{6.14}
$$

Examples for the above six scenarios are as follows: $s_{0,2} = 1$, $s_{1,-3} = s_{1,-2} = 4$, $s_{2,-1} = s_{2,0} + s_{1,-1} = 10$, $s_{3,0} = s_{2,1} + s_{2,-1} = 20$, $s_{3,1} = s_{2,2} + s_{2,0} = 34$, and $s_{3,3} = s_{3,2} = 14$. Based on the maximal procedure criteria, all feasible sequences are assigned with the same probability. Therefore, $p_{w,d,A}$ and $p_{w,d,B}$ are proportional to the number of sequences passing through the current node (n_A, n_B) to the next node $(n_A + 1, n_B)$ or $(n_A, n_B + 1)$, respectively. For two-arm balanced allocation trials, we obtain these conditional allocation probabilities:

$$
p_{w,d,A} = \begin{cases}
1 & (w = 0 \text{ and } d < 0) \text{ or } (d = -\text{MTI}) \\
\dfrac{s_{w,d+1}}{s_{w,d+1} + s_{w-1,d-1}} & (w > 0) \text{ and } (-\text{MTI} < d < 0) \\
\dfrac{s_{w-1,1}}{s_{w-1,-1} + s_{w-1,1}} = \dfrac{1}{2} & d = 0 \\
\dfrac{s_{w-1,d+1}}{s_{w,d-1} + s_{w-1,d+1}} & (w > 0) \text{ and } (0 < d < \text{MTI}) \\
0 & (w = 0 \text{ and } d > 0) \text{ or } (d = \text{MTI})
\end{cases}
\tag{6.15}
$$

Table 6.2 shows the results for two-arm trials with balanced allocations and MTI of 3 and 4, respectively. Because of the symmetric nature for balanced allocations, there are $s_{w,d} = s_{w,-d}$ and $p_{w,d,A} = p_{w,-d,B} = 1 - p_{w,-d,A}$ for $0 < d \leq \text{MTI}$.

TABLE 6.2

Number of Feasible Sequences and Conditional Allocation Probabilities for Maximal Procedure in Two-Arm Equal Allocation Trials

MTI	w	$s_{w,A}$					$p_{w,A}$				
		$d=0$	$d=1$	$d=2$	$d=3$	$d=4$	$d=0$	$d=1$	$d=2$	$d=3$	$d=4$
3	0		1	1	1			0	0	0	
	1	2	3	4	4		0.5	0.3333	0.2500	0	
	2	6	10	14	14		0.5	0.4000	0.2857	0	
	3	20	34	48	48		0.5	0.4118	0.2917	0	
	4	68	116	164	164		0.5	0.4138	0.2927	0	
	5	232	396	560	560		0.5	0.4141	0.2929	0	
	6	792	1352	1912	1912		0.5	0.4142	0.2929	0	
	7	2704	4616	6528	6528		0.5	0.4142	0.2929	0	
	8	9232	15,760	22,288	22,288		0.5	0.4142	0.2929	0	
4	0		1	1	1	1		0	0	0	0
	1	2	3	4	5	5	0.5	0.3333	0.2500	0.2000	0
	2	6	10	15	20	20	0.5	0.4000	0.3333	0.2500	0
	3	20	35	55	75	75	0.5	0.4286	0.3636	0.2667	0
	4	70	125	200	275	275	0.5	0.4400	0.3750	0.2727	0
	5	250	450	725	1000	1000	0.5	0.4444	0.3793	0.2750	0
	6	900	1625	2625	3625	3625	0.5	0.4462	0.3810	0.2759	0
	7	3250	5875	9500	13,125	13,125	0.5	0.4468	0.3816	0.2762	0
	8	11,750	21,250	34,375	47,500	47,500	0.5	0.4471	0.3818	0.2763	0
	9	42,500	76,875	123,475	171,875	171,875	0.5	0.4472	0.3819	0.2764	0
	10	153,750	278,125	450,000	621,875	621,875	0.5	0.4472	0.3819	0.2764	0

Note: d = Treatment imbalance level; $p_{w,d,A}$ = Conditional allocation probability for assigning the next subject to treatment A; $s_{w,d,A}$ = Number of feasible sequences to the balanced terminal node. w = Allocation wave number.

Therefore, only results for nonnegative d values are listed in Table 6.2. It is clear that the conditional allocation probability $p_{w,d,A}$ changes as the allocation wave w changes. It is important to indicate that $s_{w,d}$ and $p_{w,d,A}$ data listed in Table 6.2 are not affected by the allocation sequence length, because they are fully determined backward from the terminal nodes. As long as the sequence includes node$_{w,d}$, the corresponding $s_{w,d}$ and $p_{w,d,A}$ remain the same.

With minor modification, the strategy of using the allocation wave w and treatment imbalance d to calculate the conditional allocation probability for the maximal procedure applies to unequal allocations. Consider a two-arm trial with allocation ratio $r_A : r_B$, with both r_A and r_B being positive integers. The allocation wave number w and imbalance level d are defined as below:

$$w = \min\{\text{int}[N/(r_A + r_B) - n_A/r_A], \text{int}[N/(r_A + r_B) - n_B/r_B]\}$$

$$d = r_B n_A - r_A n_B$$

Figure 6.2 depicts the allocation waves for a trial with allocation $r_A : r_B = 3 : 2$, sequence length $N = 20$, and MTI $= 6$. For example, there are 444 allocation sequences from node$_{2,1}$ to terminal node$_{0,0}$, 238 pass through node$_{2,3}$, and 206 pass through node$_{2,-2}$ when the current subject is assigned to arms A and B, respectively.

Based on the definition of feasible allocation sequence and maximal procedure's equal probability sequence feature, the conditional allocation

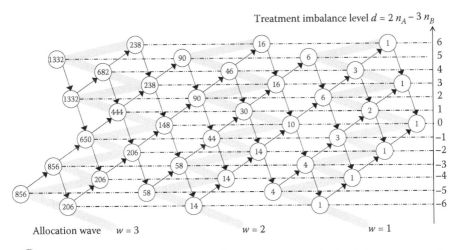

FIGURE 6.2
Allocation wave and treatment imbalance for maximal procedure sequences (two-arm, 3:2 allocation, MTI = 6, sequence length N infinitely large).

probability assigning the current subject to arm A is proportional to the number of feasible allocation sequences from the current node to the immediate upright connected nodes. For example, $p_{2,1,A} = 238 / 444 = 0.5360$. Table 6.3 lists the number of feasible allocation sequences and conditional allocation probabilities for five allocation waves.

As shown in Table 6.3, when MTI is enforced in trials with unequal allocation, a deterministic assignment for a treatment arm is used when otherwise the imbalance will go beyond the MTI. For example, at $node_{1,5}$, the imbalance has not reached the MTI yet. However, a treatment assignment to arm A will yield an imbalance of 7, which exceeds the MTI of 6. Therefore, the subject will be assigned to arm B deterministically. Similar scenarios occur at $node_{w,d}$ when $w > 0$ and $d = -4, -5$, or 5.

Figures 6.1 and 6.2 also indicate that the number of allocation nodes between two consecutive waves with the same imbalance level equals the minimal number of treatment assignments in a balanced set. For example, in Figure 6.1 with 1:1 allocation, two treatment assignments are made from $node_{w+1,d}$ to $node_{w,d}$. In Figure 6.2 with 2:3 allocation, it takes five treatment assignments.

The entries listed in Tables 6.2 and 6.3 show that with a given allocation wave w, the conditional allocation probability $p_{w,d,A}$ varies as treatment imbalance level k changes. However, for a given imbalance level d, the value of $p_{w,d,A}$ stabilizes after a few allocation waves. When the sequence length is large enough, it is expected that $p_{w,d,A}$ converges to its asymptotic value, denoted as $p_{d,A}$, which is no longer dependent on the allocation wave number w. In fact, it is solely determined by the treatment imbalance level d and the MTI. This leads to the asymptotic maximal procedure. Using the asymptotic $p_{d,A}$ as the conditional allocation probability for *all* treatment assignments, we ensure that the resulting asymptotic maximal procedure will have the Markovian property, and that its implementation will become much more straightforward.

For two-arm balanced trials, the value of $p_{d,A}$ can be obtained mathematically.[8] For unbalanced allocations, the calculation of the analytical results of conditional allocation probabilities will be more complex. However, as shown in both Tables 6.2 and 6.3, $p_{w,d,A}$ converges quickly after a few allocation waves. Therefore, it is sufficient to use the value of $p_{w,d,A}$ with large allocation wave number w as its asymptotic value. Table 6.4 provides the conditional allocation probabilities of the asymptotic maximal procedure for two-arm trials with commonly used unequal allocation ratios and MTI values. With the data in Table 6.4, the asymptotic maximal procedure can be easily implemented.

6.2.7 Comparison of Conditional Allocation Probabilities

Different randomization designs may have been illustrated by different models. If two designs have the same conditional allocation probability in

TABLE 6.3

Number of Feasible Sequences and Conditional Allocation Probabilities for Maximal Procedure in Two-Arm Trials with 3:2 Allocation and MTI = 6

Wave													
						Treatment Imbalance Level $d = 2n_A - 3n_B$							
w	-6	-5	-4	-3	-2	-1	0	1	2	3	4	5	6
S_{wd}													
0	1	4	1	4	1	3	1	2	6	1	3	6	1
1	14	58	14	58	14	44	10	30	90	16	46	90	16
2	206	856	206	856	206	650	148	444	1332	238	682	1332	238
3	3044	12,652	3044	12,652	3044	9608	2188	6564	19,692	3520	10,084	19,692	3520
4	44,996	187,024	44,996	187,024	44,996	142,028	32,344	97,032	291,096	52,036	149,068	291,096	52,036
p_{wdA}													
0	1	1	1	0.75	1	0.6667	0.6	0.5	0.5	0	0.3333	0	0
1	1	1	1	0.7586	0.7143	0.6818	0.6081	0.5333	0.5111	0.3750	0.3478	0	0
2	1	1	1	0.7593	0.7184	0.6831	0.6088	0.5360	0.5120	0.3782	0.3490	0	0
3	1	1	1	0.7594	0.7188	0.6832	0.6088	0.5363	0.5121	0.3784	0.3491	0	0
4	1	1	1	0.7594	0.7188	0.6832	0.6088	0.5363	0.5121	0.3784	0.3491	0	0

TABLE 6.4

Conditional Allocation Probability and Allocation Randomness of Asymptotic Maximal Procedure

Allocation	1:1 ($d = n_A - n_B$)				2:1 ($d = n_A - 2n_B$)				3:2 ($d = 2n_A - 3n_B$)			
MTI	2	3	4	5	3	4	5	6	5	6	7	8
Treatment imbalance level d												
−8												1
−7											1	1
−6								1		1	1	1
−5				1			1	1	1	1	1	0.78878
−4			1	0.73205		1	1	0.82843	1	1	0.77740	0.75252
−3		1	0.72361	0.63397	1	1	0.81913	0.79289	1	0.75941	0.73905	0.73221
−2	0.66667	0.70711	0.61803	0.57735	0.76759	0.80254	0.77920	0.73880	0.72871	0.71882	0.71366	0.67113
−1	0.5	0.58578	0.55279	0.53590	0.69722	0.75396	0.71663	0.70711	0.68614	0.68319	0.64690	0.63427
0	0.33333	0.5	0.5	0.5	0.62925	0.67368	0.67609	0.67157	0.62772	0.60884	0.59877	0.61348
1	0	0.41421	0.44721	0.46410	0.41080	0.61125	0.62670	0.63870	0.54257	0.53628	0.57568	0.56609
2		0.29289	0.38197	0.42265	0	0.52049	0.57313	0.6	0.40693	0.51209	0.50478	0.53868
3		0	0.27639	0.36603		0.37936	0.49644	0.55228	0.38564	0.37843	0.47397	0.48558
4			0	0.26795		0	0.36433	0.48223	0	0.34908	0.36535	0.45681
5				0			0	0.35579	0	0	0.33545	0.35009
6								0		0	0	0.32686
7											0	0
8												0
Deterministic assignments[a]	16.34%	7.16%	3.61%	2.08%	21.50%	11.94%	7.11%	4.57%	23.85%	16.10%	11.33%	8.30%
Correct guess probability[b]	66.49%	62.29%	59.75%	58.04%	68.85%	66.69%	63.72%	61.77%	69.80%	67.32%	65.04%	63.38%

Note: Allocation randomness are evaluated with sample size = 100, simulation = 50,000 per scenario.

[a] Conditional allocation probability p_A equals 0 or 1.

[b] Guess the next assignment being the least represented arm based on the target allocation.

the format Equation 6.2, they will be identical. If the conditional allocation probabilities of two randomization designs are close to each other, so will be their performances. Table 6.5 lists the conditional allocation probabilities for the permuted block design, the big stick design, the biased coin design with imbalance tolerance, and the asymptotic maximal procedure. The original maximal procedure is not included, because its conditional allocation probability is very close to that of the asymptotic maximal procedure when the allocation sequence length is not too small.

The permuted block design does not have the conditional allocation probability in the format of Equation 6.2. For a given imbalance level, it may have more than one treatment allocation status, with each having different conditional allocation probabilities. For example, for a two-arm trial with unequal allocation 2:1, let $d = n_A - 2n_B$ be the treatment imbalance, and maximal tolerated imbalance MTI = 4, corresponding to a block size of 6. Treatment distributions (1,2) and (3,2) both have $d = -1$, but have $p_A = 0.75$ and $p_A = 1.0$, respectively.

Table 6.5 reveals that when the MTI is reached, all MTI procedures use deterministic assignments; when perfect balance is achieved, all MTI procedures use complete randomization. Therefore, the true differences among these MTI procedures come from the mechanism for the handling of the

TABLE 6.5

Comparison of Conditional Allocation Probabilities among MTI Procedures (Probability of Assigning the Current Subject to Treatment Arm *A*)

| | $r_1{:}r_2 = 1{:}1;\, d = n_A - n_B;\, \text{MTI} = 3$ | | | | Imbalance | $r_1{:}r_2 = 2{:}1;\, d = n_A - 2n_B;$ MTI = 4 | | |
PBD*	BUD	BSD	BCDWIT ($p_{bc} = 0.75$)	AMP	d	PBD*	BUD	AMP
					−4	1.0	1.0	1.0
1.0	1.0	1.0	1.0	1.0	−3	1.0	1.0	1.0
0.75 or 1.0	0.75	0.5	0.75	0.707	−2	0.8 or 1.0	0.8	0.803
0.6 or 0.667 or 1.0	0.6	0.5	0.75	0.585	−1	0.75 or 1.0	0.75	0.754
0.5	0.5	0.5	0.5	0.5	0	0.667	0.667	0.674
0.4 or 0.333 or 0	0.4	0.5	0.25	0.414	1	0.6 or 0.5	0.6	0.611
0.25 or 0	0.25	0.5	0.25	0.293	2	0.5 or 0	0.5	0.520
0	0	0	0	0	3	0.333 or 0	0.333	0.379
					4	0	0	0

Note: AMP, asymptotic maximal procedure; BCDWIT, biased coin design with imbalance tolerance; BSD, big stick design; BUD, block urn design; PBD, permuted block design.
*Permuted block design may have multiple nodes with the same imbalance value. With 1:1 allocation and MTI = 3, nodes (3,2), (2,1), and (1,0) all have $d = 1$. But p_A is equal to 0.6, 0.667, and 1.0, respectively.

conditional allocation probability when $0 < |d| < $ MTI. The big stick design takes no action and keeps using the complete randomization until the MTI is reached. Chen's procedure uses a fixed biased coin probability, no matter how small or large the imbalance is. The block urn design and the asymptotic maximal procedure are equipped with an imbalance-adaptive biased coin probability. The advantage of this mechanism becomes clearer when the MTI increases.

It is worth noticing that, under an unequal allocation, the asymptotic maximal procedure has a conditional allocation probability slightly different from the target allocation probability when the imbalance is zero. This is a cost paid to have the equal allocation sequence probability.[8] It creates a scenario in which the unconditional allocation probability is not preserved.[19] However, consider the actual amount of the difference in the conditional allocation probability; its potential consequence of selection bias is practically negligible.

6.3 The Treatment Allocation Predictability

Statistical properties of randomization designs are usually evaluated based on treatment imbalance and allocation randomness.[17] Because all MTI procedures contain treatment imbalances within the MTI, the allocation randomness becomes the focus, which can be assessed by the proportion of deterministic assignment and the correct guess probability originally defined by Blackwell and Hodges.[20]

For the permuted block design, a generic formula for the proportion of deterministic assignment was given by Zhao and Weng.[21] For an m-arm trial with allocation $r_1 : r_2 : \cdots : r_m$ and block size of $b = \lambda \sum_{j=1}^{m} r_j$, the proportion of deterministic assignments under the permuted block design is

$\frac{1}{b} \sum_{j=1}^{m} \frac{\lambda r_j}{b - \lambda r_j + 1}$. For example, for two-arm trials with 2:1 allocation and a

block size of $b = 2(2 + 1) = 6$, the permuted block design has 28.9% deterministic assignment. For two-arm equal allocation trials, this formula

is reduced to $\frac{1}{b/2 + 1}$. With a block size of 4, 6, or 8, the permuted block

design has a deterministic assignment of 33.3%, 25%, or 20%, respectively. The high proportion of deterministic assignment associated with the permuted block design has caught serious attention from researchers, because of the potential risk of selection bias.[4,16,18,22-25] The permuted block design is widely used for many good reasons,[16] such as the consistent imbalance control, the generalizability to multi-arm and unequal allocation trials, the consistent unconditional allocation probability, and perhaps, most importantly, the simplicity for implementation, with or without a computer program. However, it is the most vulnerable randomization design to selection bias, especially when the perfect treatment masking and allocation concealment are not available.

The block urn design, the maximal procedure, and the asymptotic maximal procedure removed the constraint of the enforced block-end balance, allowed more allocation sequences, and therefore reduced the allocation predictability. Figure 6.3 demonstrates the differences of allocation sequence space for the permuted block design, the block urn design, and the maximal procedure.

When the minimal block size is used, that is, $\lambda = 1$, the permuted block design and the block urn design have the same allocation sequence space. When $\lambda > 1$, the block urn design has more allocation space. Under equal allocation, the maximal procedure, the asymptotic maximal procedure, and the block urn design have the same allocation space. However, under unequal allocation, the maximal procedure and the asymptotic maximal procedure have more allocation space than the block urn design does.

The proportion of deterministic assignments and the correct guess probability of the big stick design for two-arm equal allocation trials were given by Kundt[26] and Chen[12] as $(2MTI)^{-1}$ and $1/2 + (4MTI)^{-1}$, respectively.

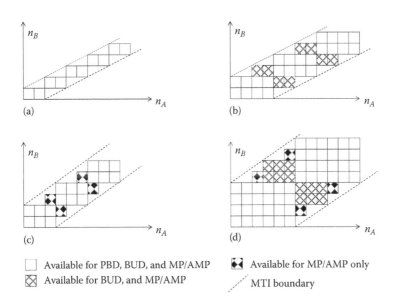

□ Available for PBD, BUD, and MP/AMP   ⬙ Available for MP/AMP only
⊠ Available for BUD, and MP/AMP   ⋰ MTI boundary

FIGURE 6.3
Comparison of feasible allocation sequences. (a) 2:1 allocation, $\lambda = 1$. (b) 2:1 allocation, $\lambda = 2$. (c) 3:2 allocation, $\lambda = 1$. (d) 3:2 allocation, $\lambda = 2$.

Figure 6.4a and b compares six randomization designs with 1:1 allocation, MTI = 3 (equivalent to block size 6 in permuted block design), and sample size varying from 10 to 100. The permuted block design has both the highest proportion of deterministic assignments and the highest correct guess probability; when the sample size is a multiple of the block size, the allocation randomness is worsened because of the enforced perfect balance at the end of each block. Chen's BCDWIT procedure uses a fixed biased coin probability to manage two competing demands. A higher biased coin probability will decrease the proportion of deterministic assignments at the cost of a higher correct guess probability, and a lower biased coin probability can reduce the correct guess probability at the cost of a higher proportion of deterministic assignments. When the biased coin probability is reduced to 0.5, Chen's BCDWIT procedure is identical to the big stick design, which has the lowest correct guess probability and a medium proportion of deterministic assignments. The maximal procedure performs better when the sample size is larger. Compared to the block urn design, the asymptotic maximal procedure has a slightly lower correct guess probability and a slightly higher proportion of deterministic assignments. Both perform better overall than the big stick design and Chen's BCDWIT procedure.

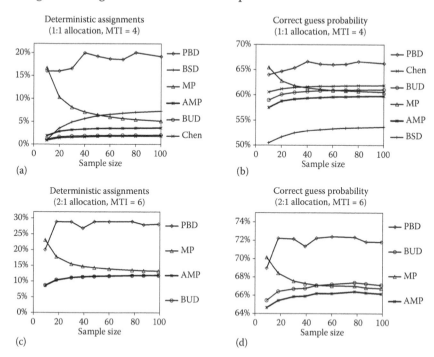

FIGURE 6.4
Comparison of allocation randomness. PBD, permuted block design; BSD, big stick design; Chen, Chen's procedure with biased coin probability of 0.65; BUD, block urn design; MP, maximal procedure; AMP, asymptotic maximal procedure.

Figure 6.4c and d shows the simulation results for unequal allocation 2:1 with MTI of 6 (equivalent to block size of 9 in permuted block design) and sample size varying from 9 to 99. Once again, the permuted block design has the worst performance. The maximal procedure has good performance when the sample size is larger than 40. The block urn design and the asymptotic maximal procedure have the same low proportion of deterministic assignments. The asymptotic maximal procedure has a lower correct guess probability. The big stick design and Chen's procedure are not included in this comparison as both were not defined for unequal allocations.

6.4 Implementation of MTI Procedures

The old pre-generated permuted block design sequences stratified by site using concealed envelopes is outdated[4] and should be replaced by better alternative randomization designs and more secure implementation methods. Ideally, a computerized subject randomization module is integrated with the clinical trial management and electronic data capture system, so that a treatment assignment is generated at real time whenever an eligible subject is ready for randomization. This strategy eliminates the risk of treatment assignment concealment failure and provides options for imbalance-adaptive randomization within baseline covariate strata across clinical sites. With the information of current treatment distribution and the conditional allocation probability formula of the selected randomization algorithm, the treatment randomization can be carried out by comparing the conditional allocation probability to a random number with the uniform distribution in (0,1). An optional pre-generated random number list could be used, so that the computerized subject randomization system can be tested and validated properly. This approach works only when the explicit conditional allocation probability formula is available.

6.5 Summary

Restricted randomization designs with maximal tolerated imbalance are applied in clinical trials to consistently control the treatment imbalances for the purpose of preventing potential chronological biases. Behind any gain in treatment balance, there is a cost in allocation randomness. However, for the same amount of gain in treatment balance defined by the MTI, different randomization designs pay different costs in the allocation randomness.

The obvious inferiority of the permuted block design is the consequence of the enforced balance at the end of each block. For two-arm equal allocation trials, the block urn design, the asymptotic maximal procedure, and the big stick design are better alternatives to the permuted block design. For multi-arm trials and trials with unequal allocations, the block urn design and the asymptotic maximal procedure are recommended.

References

1. Berger VW, Ivanova A, Deloria-Knoll M, Minimizing predictability while retaining balance through the use of less restrictive randomization procedures. *Stat Med* 2003; **22**:3017–3028.
2. Salama I, Ivanova A, Qaqish B, Efficient generation of constrained block allocation sequences. *Stat Med* 2008; **27**(9):1421–1428.
3. Rosenberger WF, Lachin JM, *Randomization in Clinical Trials Theory and Practice*, Wiley-Interscience, New York, 2002.
4. Zhao W, Selection bias, allocation concealment and randomization design in clinical trials. *Contemp Clin Trials* 2013; **36**:263–265.
5. Soares JF, Wu CF, Some restricted randomization rules in sequential designs. *Commun Stat Theory Methods* 1983; **12**:2017–2034.
6. Chen YP, Biased coin design with imbalance tolerance. *Commun Stat Stochastic Models* 1999; **15**:953–975.
7. Zhao W, Weng Y, Block urn design—A new randomization algorithm for sequential trials with two or more treatments and balanced or unbalanced allocation. *Contemp Clin Trials* 2011; **32**:953–961.
8. Zhao W, Berger VW, Yu Z, The asymptotic maximal procedure for subject randomization in clinical trials. *Stat Methods Med Res.* DOI: 10.1177/0962280216677107. In print.
9. Efron B, Forcing a sequential experiment to be balanced. *Biometrika* 1971; **58**:403–417.
10. Wei LJ, A class of designs for sequential clinical trials. *J Am Stat Assoc* 1977; **72**:382–386.
11. Smith RL, Sequential treatment allocation using biased coin designs. *J R Stat Soc Series B Methodol* 1984; **46**:519–543.
12. Chen YP, Which design is better? Ehrenfest urn versus biased coin. *Adv Appl Probab* 2000; **32**:738–749.
13. Antognini AB. Extensions of Ehrenfest's urn designs for comparing two treatments. *mODa 7 Contrib Stat* 2004; 21–28.
14. Hill AB, The clinical trial. *Br Med Bull* 1951; **71**:278–282.
15. McEntegart DJ, The pursuit of balance using stratified and dynamic randomization techniques: An overview, *Drug Inf J* 2003; **37**:293–308.
16. Zhao W, A better alternative to the inferior permuted block design is not necessarily complex, *Stat Med* 2016; **35**:1736–1738.

17. Zhao W, Weng Y, Wu Q, Palesch Y, Quantitative comparison of randomization designs in sequential clinical trials based on treatment balance and allocation randomness. *Pharm Stat* 2012; **11**:39–48.
18. Zhao W, A better alternative to stratified permuted block design for subject randomization in clinical trials. *Stat Med* 2014; **33**:5329–5348.
19. Kuznetsova OM, Tymofyeyev Y, Preserving the allocation ratio at every allocation with biased coin randomization and minimization in studies with unequal allocation. *Stat Med* 2012; **31**(8):701–723.
20. Blackwell D, Hodges JL, Design for the control of selection bias. *Ann Math Stat* 1957; **28**:449–460.
21. Zhao W, Weng Y, A simplified formula for quantification of the probability of deterministic assignment in permuted block randomization. *J Stat Plan Inference* 2011; **141**(1):474–478.
22. Berger VW, *Selection Bias and Covariate Imbalances in Randomized Clinical Trials*, John Wiley & Sons, 2005.
23. Berger VW, Do not use blocked randomization. *Headache* 2006A; **46**(2):343.
24. Berger VW, Misguided precedent is not a reason to use permuted blocks. *Headache* 2006B; **46**(7):1210–1212.
25. Berger VW, Varying block sizes does not conceal the allocation. *Crit Care* 2006C; **21**(2):229.
26. Kundt G, A new proposal for setting parameter values in restricted randomization methods. *Methods Inf Med* 2007; **46**:440–449.

7

Evaluating the Evaluation

Adriana C. Burgos and Ross J. Kusmick

CONTENTS

7.1 Introduction 83
7.2 Our Focus 83
7.3 Classification of Risk of Bias 84
7.4 Why Are These Studies Flawed? 84
7.5 Consequences to Using Suboptimal Studies 86
References 87

7.1 Introduction

The purpose of randomization is to prevent bias from being introduced into a study. When the appropriate method of randomization is used and all the information concerning the process of randomization is given, the risk for bias decreases and the validity of the study increases. In this article, we will evaluate the assessment of risk of bias presented by Elangovan et al. Elangovan et al. [1] evaluated 17 periodontal regeneration trials to determine their risk of bias and concluded that 7 of these 17, representing more than 40% of the trials, had a high risk for bias. Even though this high percentage already calls into question the legitimacy of these trials taken as a whole, the assessment still seems to be too forgiving, as we intend to demonstrate.

7.2 Our Focus

We draw attention to one key component of trial quality, namely, the use of an appropriate randomization method, which we acknowledge is rather elusive in *all* trials, and not just in these 17. It has already been established that the most common methods of randomization, which are permuted blocks and minimization, are flawed procedures that do not guard against selection bias. It is worth noting that not only have permuted blocks been repeatedly criticized in the literature (see, for example, Refs. [2] and [3], and additional references contained therein), but in fact these criticisms have never been refuted. Therefore, any trial that either (1) used one of these flawed randomization methods or (2) did not specify how randomization was conducted would be at a high risk for selection bias, above and beyond the risks of other biases.

7.3 Classification of Risk of Bias

So how do the 17 trials stack up with regard to the randomization method? We note that of the 17 trials, the authors rated 7 as being at high risk of bias, 5 being of low risk, and the remaining 5 being classified as unclear risk of bias. This, of course, is an overall assessment and does not focus exclusively on randomization and allocation concealment, as we intend to do. Before we do, however, we note the fact that "unclear risk of bias" is somewhat of an oxymoron. In the assessment by Elangovan et al., "unclear risk" is the assigned grade if sequence generation, allocation concealment, and/or masking of participants is missing or is hard to interpret, or for incomplete outcome data or selective outcome reporting, although the very word "risk" already implies the potential, or the possibility, rather than the certainty. Therefore, we do not need to be certain that a bias exists to rate a trial as being at high *risk* for bias. Whenever we cannot rule bias out, as is understood with an unclear risk of bias, it is tantamount to a high risk of bias [4–7]. This being the case, of the 17 trials conducted, we interpret the findings as 12 being at high risk of bias and only 5 being at low risk. Our aim is to demonstrate that even this already damning finding is still too lenient, so we focus on only the five trials that are ostensibly at low risk.

For the sake of clarity and consistency, we shall adhere to the numbering of the 17 trials used by the original authors. The five trials considered to be at low risk for bias were numbers 4 [8], 5 [9], 11 [10], 13 [11], and 16 [12]. This is how we shall refer to them in the remainder of this article.

7.4 Why Are These Studies Flawed?

Article number 4 included 34 patients with three or more intraosseous defects [8]. A defect number was assigned to each of the 108 possible defects studied. The defect number was then randomly allocated to one of the three treatment regimens. It is important to note that the unit of randomization was the defect, and not the patient, an irony on the word "defect" as this in and of itself already constitutes defective randomization. But beyond this, all we are told about how the randomization was conducted is that a custom-made, computer-generated table determined this assignment. No information was provided regarding *how* this table was prepared.

The authors of article number 5 randomly assigned the patients to one of the two treatment groups [9]. As seen in article number 4, the assignment was also conducted with a custom-made computer-generated table. The authors did mention that, for the purpose of concealment, patients were given an opaque envelope, which indicated the assigned treatment group. We are told absolutely nothing regarding how randomization was conducted or how the allocation was concealed. Stating that patients were given an opaque envelope is not enough. It is yet another study with an inadequate description of the randomization process.

Article 11 used, what the authors referred to as, "sequence generation" and "allocation concealment" [10]. A total of 24 small sheets of paper were folded and mixed into a box: 12 pertaining to the treatment and the remaining 12 pertaining to the control. Before the initial surgery, the patients selected a sheet of paper from the box. Written on the chosen sheet of paper was the treatment group assigned to the patient. There was, at least in theory, no way to predict the therapy that the patient would receive, or so the authors claimed. Once again, no additional specifications were made about the process of neither randomization nor allocation concealment.

Article number 13 randomly assigned patients to a treatment group using block randomization, with a block size of four [11]. The only information provided was that "randomization was performed by the biostatistician, who generated a computer sequence of numbers using a block randomization (four-unit block size)." Again, no details were provided about how randomization was achieved. Blocks have been shown to lead to excessive allocation prediction, consequently potentially introducing selection bias. Article 13 is considered low risk, but this could be disputed if the integrity of the trial was compromised. The Berger–Exner test [2] can provide information to conclude whether or not selection bias is present and offers a way to correct it.

For the 16th article, treatment was determined during the surgical procedure by a computer-generated list [12]; this is all we are told. How the randomization was conducted was completely omitted, just as we observed with some of the other trials.

We have already noted that information insufficient to rule a bias out is tantamount to a high risk for that bias. Therefore, the authors were too lenient in scoring articles number 4, 5, 11, 13, and 16 as low risk for bias. Since we do not know how they randomized, clearly we cannot discard high risk for selection bias, and therefore, they are also at a high risk for overall bias (including selection bias). These five articles did not provide sufficient information on the process of randomization, besides them very loosely using the term "randomization."

7.5 Consequences to Using Suboptimal Studies

Why is this so important? The only way for a reader to determine if the study was properly executed is in the details used in its description. If the details are omitted, there is no way to verify the results of the study. Results that are not verifiable and not reproducible should not be used. Just using the word "randomization" or using an opaque envelope does not constitute a proper study. The reader must be able to verify that, at every step, every measure from start to finish was taken to ensure allocation concealment. The process of randomization is of utmost importance because "differences between intervention and control groups should behave like differences between two random samples from the population" [13]. Studies are used to determine the benefit of one treatment in comparison to a control, but if there are possible biases or questionable results, owing to faulty randomization or allocation concealment, then the study itself cannot be trusted. A study done by Moher et al. [14] randomly selected 11 meta-analyses and replicated their findings using the published data in the primary studies. What they found was that when different methods of quality score integration were used, the perceived benefit of the intervention could be changed, especially when the study was found to be of low quality.

The significance in holding trials at a higher standard, in terms of risk of bias and quality of a study, is in the ramifications the flawed study may have on treatment and overall public health. Of the 17 trials assessed by Elangovan et al., they categorized only 7 as being at a high risk for bias. However, we concluded that the studies described as having an unclear risk for bias as well as those of low risk should be considered high risk as seen in Table 7.1. The repercussions for using studies with a high risk for bias are

TABLE 7.1

Flaws Encountered in Trials Initially Considered to Have a Low Risk of Bias, by Elangovan et al., as well as Our Classification of the Aforementioned Trials

Trials Initially Considered Low Risk of Bias by Elangovan et al.	Flaw Encountered	Our Classification of Risk of Bias
4	Defect used as unit of randomization instead of patient and study lacks specification on how randomization was conducted	High risk
5	Study lacks specification on how randomization was conducted and allocation concealment	High risk
11	More information needed on allocation concealment	High risk
13	Block randomization in itself is flawed and may present selection bias	High risk
16	Study lacks specification on how randomization was conducted	High risk

very important and investigators as well as publishers should ensure superior quality work. We speculate that the risks may be higher when the trial is conducted by researchers who are so cavalier as to use amply discredited randomization methods, not realizing or not caring about the impact it can have on their study or on public health. The overall risk of bias is, therefore, almost certainly higher (and not lower) than the risk of just the single bias we consider here. All it takes is one bias to completely invalidate a trial, even if that trial was perfect in every other way [15].

References

1. Elangovan S, Prakasam S, Gajendrareddy P, Allareddy V (2016). A risk of bias assessment of randomized controlled trials (RCTs) on periodontal regeneration published in 2013. *J Evid Based Dent Pract* 16, 30–40.
2. Berger VW, Bejleri K, Agnor R (2015). Comparing MTI randomization procedures to blocked randomization. *Statistics in Medicine*.
3. Berger VW (2005). *Selection Bias and Covariate Imbalances in Randomized Clinical Trials*. Chichester: John Wiley & Sons.
4. http://handbook.cochrane.org/chapter_8/table_8_5_d_criteria_for_judging_risk _of_bias_in_the_risk_of.htm
5. Berger VW (2012). Conservative handling of missing data. *Contemp Clin Trials* 33, 460.
6. Berger VW (2012). Conservative handling of missing information. *J Clin Epidemiol* 65, 1237–1238.
7. Berger VW (2006). Missing data should be more heartily penalized. *J Clin Epidemiol* 59(7), 759–760.
8. De Leonardis D, Paolantonio M (2013). Enamel matrix derivative, alone or associated with a synthetic bone substitute, in the treatment of 1- to 2-wall periodontal defects. *J Periodontol* 84(4), 444–555.
9. Di Tullio M, Femminella B, Pilloni A, Romano L, D'Arcangelo C, De Ninis P, Paolantonio M (2013). Treatment of supra-alveolar-type defects by simplified papilla preservation technique for access flap surgery with or without enamel matrix proteins. *J Periodontol* 84(8), 1100–1110.
10. Jenabian N, Haghanifar S, Maboudi A, Bijani A (2013). Clinical and radiographic evaluation of Bio-Gen with biocollagen compared with Bio-Gen with connective tissue in the treatment of class II furcation defects: A randomized clinical trial. *J Appl Oral Sci.* 21(5), 422–429.
11. Mishra A, Avula H, Pathakota KR, Avula J (2013). Efficacy of modified minimally invasive surgical technique in the treatment of human intrabony defects with or without use of rhPDGF-BB gel: A randomized controlled trial. *J Clin Periodontol* 40, 172–179.
12. Peres MFS, Ribeiro EDP, Casarin RCV, Ruiz KGS, Junior FHN, Sallum EA, Casati MZ (2013). Hydroxyapatite/b-tricalcium phosphate and enamel matrix derivative for treatment of proximal class II furcation defects: A randomized clinical trial. *J Clin Periodontol* 40, 252–259.

13. Viera AJ, Bangdiwala SI (2007). Eliminating bias in randomized controlled trials: Importance of allocation concealment and masking. *Fam Med* 39(2), 132–137.
14. Moher D, Jones A, Cook DJ, Jadad AR, Moher M, Tugwell P, Klassen TP (1998). Does quality of reports of randomised trials affect estimates of intervention efficacy reported in meta-analyses? *Lancet* 352(9128), 609–613.
15. Palys K, Berger VW, Alperson S (2013). Trial quality checklists: On the need to multiply (not add) scores. *Clin Oral Investig* 17, 1789–1790.

8

Selection Bias in Studies with Unequal Allocation

Olga M. Kuznetsova

CONTENTS

8.1 Introduction 89
8.2 The Blackwell–Hodges Model for Selection Bias in Studies
    with Equal and Unequal Allocation 91
    8.2.1 Elements of the Selection Bias in the Blackwell–Hodges
           Framework 91
    8.2.2 Expected Bias Factor 95
8.3 Directional Guessing Strategy 97
    8.3.1 Justification of the Exact Conditional Guessing Strategy 97
    8.3.2 Justification of the Directional Guessing Strategy 100
    8.3.3 Expected Bias Factor under the Directional Strategy 105
8.4 Selection Bias with PBR 107
8.5 Directional Strategy with the Payoff Threshold 113
8.6 Selection Bias When the Investigator Assumes PBR in a Study
    with the BTR 116
8.7 Discussion 119
References 120

8.1 Introduction

In open-label single-center trials (or multi-center trials with randomization stratified by center), the investigator knows the complete sequence of treatment assignments of patients already randomized into the study. Thus, when a restricted allocation procedure is used in such a trial, the investigator might try to guess the next treatment assignment and allocate the patient he considers right for the guessed treatment. This can make the treatment groups different in composition and, thus, introduce the selection bias [1].

All restricted allocation procedures with equal or unequal allocation have a potential for selection bias, but the magnitude of the selection bias differs across the allocation procedures. For equal allocation procedures, the selection bias is well studied and comparisons between procedures are extensively described [1,2]. Under the popular Blackwell–Hodges model [3], where subjects with high expected response are assigned to one treatment and subjects with low expected response are assigned to another, the expected treatment difference due to biasing depends on the proportion of the correct guesses across the two treatment arms [1]. Equal allocation procedures with higher expected proportion of correct guesses have higher potential for selection bias. Thus, the expected proportion of correct guesses presents a good tool to compare the allocation procedures in their potential for selection bias.

However, for procedures with unequal allocation, the expected proportion of correct guesses does not translate to a measure of selection bias. Indeed, if in a study with 1:9 allocation to treatments A and B the investigator always guesses the next treatment to be B, the proportion of correct guesses will be very high—0.9. This guessing strategy, though, will not result in a selection bias as the composition of both groups will be the same. Thus, the comparisons of the expected proportions of correct guesses do not seem particularly useful for unequal allocation procedures.

For unequal allocation trials, selection bias depends on the sum of expected proportions of correct guesses derived separately for treatment groups A and B. For equal allocation, this relationship is expressed in Equation 5.1 from Ref. [1] (the part of the equation that translates the proportions into the total number of correct guesses is not applicable to unequal allocation). Rosenberger and Lachin [1] offer the reader to derive the expected bias for unequal allocation as Problem 5.1, but the derivation itself is not provided in the book. It will be provided in this chapter for completeness.

The selection bias that the investigator introduces into a study depends on the guessing strategy they employ. The selection bias is commonly calculated assuming that the investigator knows what allocation procedure is used to randomize the subjects. For a large two-arm study, the highest bias is achieved when the investigator employs the directional guessing strategy [2,4] that guesses the treatment for which the conditional allocation ratio exceeds the unconditional allocation ratio. (When the conditional allocation ratio is equal to the unconditional allocation ratio, the treatment is typically guessed in the unconditional allocation ratio.)

For many restricted equal allocation procedures, the directional strategy coincides with the convergence strategy where the treatment underrepresented so far is the next guess. The reason is that for these allocation procedures, the underrepresented treatment is assigned with a conditional probability of $>1/2$. The example of an exception is the big stick design [5] for which the two strategies differ.

Since biasing is hard work and holding back patients until the proper assignment comes along is hard to justify, especially when the enrollment is slow, the investigators might choose to attempt to bias only when the return on the biasing effort is considerable. Strategies with a payoff threshold select the best matched patients only when the increment in the expected bias due to the next allocation is large [2]. For some allocation procedures, such as

Permuted Block Randomization (PBR) [6], it means that most often, no biasing occurs when the imbalance in treatment assignments is low.

Success of the directional biasing strategies with or without the payoff threshold depends on the investigator knowing the allocation procedure used in the study. However, this is not necessarily the case. Most investigators are not aware of novel allocation procedures and always expect the PBR to be employed in the study. Thus, it is of interest to evaluate the selection bias with procedures other than PBR when the investigator expects a PBR. In particular, for unequal allocation procedures that keep the imbalance in treatment assignments low, such as the Brick Tunnel Randomization (BTR) or Wide Brick Tunnel Randomization (WBTR) [7–13], the directional strategy with a payoff threshold will lead to very few attempts to bias, and thus, lower selection bias.

In Section 8.2, the quantification of the selection bias following the Blackwell–Hodges model [3] will be provided for studies with unequal allocation. Section 8.3 will discuss guessing strategies in studies with unequal allocation and provide a justification for the directional guessing strategy. Section 8.4 will discuss the selection bias characteristics for the PBR—the unequal allocation procedure most commonly used in clinical trials. In Section 8.5, the selection bias under the guessing strategy with a payoff threshold will be derived. In Section 8.6, the selection bias in a study that uses the BTR [8] while the investigator biases the allocation assuming the PBR will be quantified. A discussion completes the chapter.

8.2 The Blackwell–Hodges Model for Selection Bias in Studies with Equal and Unequal Allocation

8.2.1 Elements of the Selection Bias in the Blackwell–Hodges Framework

While the Blackwell–Hodges [3] model for selection bias was described for equal allocation to two treatment arms, it can be used with unequal allocation as well.

Consider a study with allocation in C_1:C_2 ratio to treatments A and B, where C_1 and C_2 have no common divisors other than 1. Let us call $S = C_1 + C_2$ the block size. The allocation probabilities are $w_1 = C_1/S$ and $w_2 = C_1/S$; $w_1 + w_2 = 1$.

Suppose a study uses an allocation procedure that has the set Ω of allocation sequences $\omega = \{\omega_1,..., \omega_N\}$ of the length N. Here $\omega_i = 1$ if treatment A is assigned at the ith allocation; $\omega_i = 0$ if treatment B is assigned at the ith allocation, $i = 1,..., N$. Let us denote by $p(\omega)$ the probability with which sequence ω occurs in Ω.

We will only consider the allocation procedures that preserve the unconditional allocation ratio at every allocation that is procedures for which

$$\sum_{\omega \in \Omega} p(\omega)\omega_i = w_1 \quad \text{for all } i = 1, \dots, N.$$

Kuznetsova and Tymofyeyev call such procedures Allocation Ratio Preserving (ARP) procedures [14–16]. Failure to preserve the allocation ratio at
every allocation can result in selection and evaluation bias even in double-
blind trials [14], accidental bias, and problems with the re-randomization test
[14,17]. Thus, non-ARP procedures should be generally avoided [14–16,18].
While most equal allocation procedures are ARP procedures due to symme-
try with respect to the treatment arms, many unequal allocation procedures
do not have the ARP property [19–26], with BTR, WBTR, block urn design
[27], and the play-the-winner-type urn design [28] being rare ARP exceptions.

The allocation sequence ω is drawn from Ω for the study. While the investigator does not know the upcoming assignments, the past assignments in
this open-label study are known to the investigator. Based on the knowledge
of the past assignments, the investigator is trying to guess which assignment
is coming next.

Let us denote by $g = \{g_1, \dots, g_N\}$, $g_i = 0$ or 1, the sequence of the guesses that
the investigator made for the upcoming allocations. Here, $g_i = 1$ corresponds
to guess A and $g_i = 0$ corresponds to guess B.

For simplicity, assume that the expectation of the response variable Y is the
same for both treatments.

Under the Blackwell–Hodges model, there are two types of patients:
healthier patients with the expected response $\mu + \Delta$ and sicker patients with
the expected response $\mu - \Delta$. The investigator that wants to bias the results in
favor of treatment A allocates a healthier patient when his guess for the next
treatment is A and allocates a sicker patient when his guess is B.

For unequal allocation, the expected difference in the treatment group
means can be derived following the derivations presented in Ref. [1, p. 81]
for equal allocation.

In a study with the randomization sequence ω and the sequence of guesses
g, let us denote by N_1 and N_2 the numbers of subjects and by $\alpha(g,\omega)$ and $\beta(g,\omega)$
the proportions of correct guesses in the treatment groups A and B, respec-
tively. Then,

$$N_1 = \sum_{i=1}^{N} \omega_i$$

$$N_2 = \sum_{i=1}^{N} (1 - \omega_i)$$

$$\alpha(g,\omega) = \frac{1}{N_1} \sum_{i=1}^{N} g_i \omega_i \qquad (8.1)$$

$$\beta(g,\omega) = \frac{1}{N_2} \sum_{i=1}^{N} (1-g_i)(1-\omega_i). \qquad (8.2)$$

Then, given g and ω, the expectations of the average response in groups A and B are

$$E(\overline{Y}_A) = \alpha(g,\omega)(\mu + \Delta) + (1-\alpha(g,\omega))(\mu - \Delta) = (\mu - \Delta) + 2\alpha(g,\omega)\Delta$$

$$E(\overline{Y}_B) = (1-\beta(g,\omega))(\mu + \Delta) + \beta(g,\omega)(\mu - \Delta) = (\mu + \Delta) - 2\beta(g,\omega)\Delta.$$

Thus, the expected difference in the group averages for given g and ω is

$$D_1(g,\omega) = E(\overline{Y}_A - \overline{Y}_B \mid g,\omega) = 2\Delta(\alpha(g,\omega) + \beta(g,\omega) - 1). \qquad (8.3)$$

This is the expected bias in the treatment effect for the observed sequences g and ω. The expectation in this equation is over the distribution of the random responses.

The investigator makes a guess g_i based on the $\omega_1,\ldots,\omega_{i-1}$ known to them and following their guessing strategy. Since the strategy can involve random guesses, g_i can be thought of as a random binary function of $\omega_1,\ldots,\omega_{i-1}$: $g_i = g_i(\omega_1,\ldots,\omega_{i-1})$ with the expectation f_i. When $f_i = 1$ or $f_i = 0$, the guess g_i does not involve a random element and is deterministically A or B, respectively.

Let us denote by $D_2(\omega)$ the expected value of $D_1(g,\omega)$ over the distribution of the random guesses given the allocation sequence ω. Then,

$$D_2(\omega) = E(\overline{Y}_A - \overline{Y}_B \mid \omega) = 2\Delta(\alpha_\omega + \beta_\omega - 1), \qquad (8.4)$$

where α_ω and β_ω are the expected proportions of correct guesses in treatment groups A and B, respectively, given ω,

$$\alpha_\omega = E(\alpha(g,\omega)\mid \omega) = \frac{1}{N_1} \sum_{i=1}^{N} \omega_i f_i \qquad (8.5)$$

$$\beta_\omega = E(\beta(g,\omega)\mid \omega) = \frac{1}{N_2} \sum_{i=1}^{N} (1-\omega_i)(1-f_i). \qquad (8.6)$$

$D_2(\omega)$ is the expected selection bias in the treatment effect for the allocation sequence ω. To increase the selection bias, the investigator should try to increase the sum of the proportions of correct guesses in the two treatment groups. The extent of the bias he can introduce in this way depends on the allocation sequence ω that was generated for the study. For some allocation sequences, $D_2(\omega)$ will be high, and for other sequences, it will be low.

For example, consider an investigator who uses the convergence strategy in a study with equal allocation to treatments A and B and $N = 2n$. If ω is a sequence of alternating assignments ABAB...AB, the balance in treatment assignments will be achieved after every even allocation. Thus, the investigator will guess every odd allocation with 1:1 odds and will guess B for every even allocation. Thus, from Equations 8.5 and 8.6, $\alpha_\omega = 1/2$, $\beta_\omega = 1$, and, from Equation 8.4, $D_2(\omega) = 2\Delta \times 1/2$.

If the sequence ω is a sequence AABABA...BABB, the investigator will make the first guess at random with 1:1 odds. However, since from the second allocation on there are always more subjects allocated to A than to B, the investigator will guess B for all remaining allocations. Therefore, from Equations 8.5 and 8.6, $\alpha_\omega = 1/(2n)$, $\beta_\omega = 1$, and, from Equation 8.4, $D_2(\omega) = 2\Delta \times 1/(2n)$.

$D_2(\omega)$ reaches its maximum of 2Δ when the allocation sequence is known to the investigator, and thus all guesses are correct; this leads to $\alpha_\omega = \beta_\omega = 1$.

If the investigator makes his guesses completely at random with probabilities p_1 and p_2, $p_1 + p_2 = 1$, then $\alpha_\omega = p_1$, $\beta_\omega = p_2$, and thus from Equations 8.5 and 8.6, $\alpha_\omega + \beta_\omega - 1 = 0$ and, from Equation 8.4, $D_2(\omega) = 0$. This shows that guessing at random does not lead to the selection bias.

While the expected selection bias in a particular study depends on the allocation sequence generated for the study, the choice of the allocation procedure is influenced by the expectation D_3 of the selection bias over the set of its allocation sequences $\{\Omega, p(\omega)\}$:

$$D_3 = E(\bar{Y}_A - \bar{Y}_B) = \sum_{\omega \in \Omega} D_2(\omega)p(\omega) = 2\Delta(\alpha + \beta - 1), \tag{8.7}$$

where α and β are the expected proportions of correct guesses in the treatment groups A and B, respectively, for the given allocation procedure and the guessing strategy:

$$\alpha = \sum_{\omega \in \Omega} \alpha(\omega)p(\omega)$$

and

$$\beta = \sum_{\omega \in \Omega} \beta(\omega)p(\omega).$$

8.2.2 Expected Bias Factor

When the Blackwell–Hodges model and the expected selection bias formula (Equation 8.7) are discussed for equal allocation, the proportions of correct guesses are usually converted into the numbers of correct guesses. This is done assuming that at the end of the study, the treatment group totals are exactly equal. While in practice the allocation procedures are rarely set up to have exactly the same treatment group totals, for large N, the observed allocation ratio is indeed close to 1:1 for most restricted randomization procedures. However, this is not necessarily the case for smaller N, in particular, when N is the size of a center in a multi-center study, or the size of a stratum within a center in a study with randomization stratified by center and other factors. For example, when a PBR allocation with a permuted block size (PBS) of 8 is used in a center that enrolls four subjects, the split of the treatment group totals is not necessarily $2 + 2$, but could be with notable probabilities $3 + 1$ or $1 + 3$ (and with smaller probabilities $4 + 0$ or $0 + 4$).

The formulas for selection bias derived above are based on the proportions of the correct guesses and thus do not require any assumptions regarding the final treatment totals N_1 and N_2. In the remainder of the chapter, we will point out considerations that require such assumption.

For equal allocation and under the assumption $N_1 = N_2 = N/2$, Equation 8.7 reduces to

$$E\left(\overline{Y_A} - \overline{Y_B}\right) = 2\Delta \frac{E\left(G - \dfrac{N}{2}\right)}{N/2},$$

where G is the total number of correct guesses [1]. The numerator

$$E(F) = E(G) - \frac{N}{2}$$

is called the expected bias factor [1].
Thus,

$$E(F) = \frac{N}{2} E\left(\overline{Y_A} - \overline{Y_B}\right)/2\Delta. \tag{8.8}$$

The expected bias factor (Equation 8.8) is a part of the selection bias that depends only on the allocation procedure and the guessing strategy, but not on the expected difference in response between the healthier and sicker patients. If the allocation follows a deterministic sequence known to the investigator, the expected bias factor is $N/2$. For 1:1 complete randomization, the expected bias factor is 0. For other equal allocation procedures, when the investigator knows the procedure used in a trial, the expected bias factor is between 0 and $N/2$.

The expected bias factor is derived analytically [1] for many equal allocation procedures and derived through simulations otherwise [1]. The equal allocation procedures are compared in their potential for selection bias in an open-label study using the expected bias factor.

The concept of the expected bias factor that relies on the assumption of the final split in the treatment group counts can be introduced for procedures with unequal allocation as well. But before we do that, let us introduce the concept of the exact expected bias factor $E(F_X)$ that can be used with equal or unequal allocation and does not rely on such an assumption. Specifically, using Equation 8.8 as the definition of the exact expected bias factor, we obtain

$$E(F_X) = \frac{N}{2}(\alpha + \beta - 1). \tag{8.9}$$

Equations 8.3, 8.4, and 8.7 for the expected selection bias and Equation 8.9 for the exact expected bias factor under the Blackwell–Hodges model [3] can also be expressed through the proportions of correct and incorrect guesses.

Let us denote by $\alpha'(g,\omega)$ and $\beta'(g,\omega)$ the proportions of incorrect guesses in the treatment groups A and B, respectively:

$$\alpha'(g,\omega) = \frac{1}{N_1}\sum_{i=1}^{N}(1 - g_i)\omega_i \tag{8.10}$$

$$\beta'(g,\omega) = \frac{1}{N_2}\sum_{i=1}^{N}g_i(1 - \omega_i). \tag{8.11}$$

Then, since $\alpha(g,\omega) + \alpha'(g,\omega) = 1$ and $\beta(g,\omega) + \beta'(g,\omega) = 1$,

$$\alpha(g,\omega) + \beta(g,\omega) - 1 = (\alpha(g,\omega) - \alpha'(g,\omega) + \beta(g,\omega) - \beta'(g,\omega))/2$$

and Equation 8.3 can be written as

$$D_1(g,\omega) = E(\bar{Y}_A - \bar{Y}_B|g,\omega) = 2\Delta(\alpha(g,\omega) - \alpha'(g,\omega) + \beta(g,\omega) - \beta'(g,\omega))/2. \tag{8.12}$$

Similarly, Equation 8.4 can be written as

$$D_2(\omega) = E(\bar{Y}_A - \bar{Y}_B|\omega) = 2\Delta\left(\alpha_\omega - \alpha'_\omega + \beta_\omega - \beta'_\omega\right)/2, \tag{8.13}$$

where α'_ω and β'_ω are the expected proportions of incorrect guesses in treatment groups A and B, respectively, given ω. And finally, Equation 8.7 can be written as

$$D_3 = E(\bar{Y}_A - \bar{Y}_B) = 2\Delta(\alpha - \alpha' + \beta - \beta')/2, \tag{8.14}$$

where α' and β' are the expected proportions of incorrect guesses in the treatment groups A and B, respectively, for given allocation procedure and the guessing strategy.

From Equations 8.9 and 8.14, the exact expected bias factor can be expressed as

$$E(F_X) = \frac{N}{4}(\alpha - \alpha' + \beta - \beta'). \tag{8.15}$$

For equal allocation and assuming $N_1 = N_2 = N/2$, Equation 8.15 reverts to the expected bias factor formula (Equation 5.3) from Ref. [1]:

$$E(F) = E(H - M)/2.$$

Here, H is the total number of correct guesses and M is the total number of incorrect guesses. Although the formula is proven for the convergence strategy in Ref. [1], it works for any guessing strategy as shown above.

8.3 Directional Guessing Strategy

8.3.1 Justification of the Exact Conditional Guessing Strategy

Let us denote by $v_1(X,Y)$ and $v_2(X,Y)$ the conditional allocation probabilities of treatments A and B, respectively, after X assignments to A and Y assignments to B were made; $v_1(X, Y) + v_2(X, Y) = 1$. For example, for 7:9 PBR with a block size of 16 after 3 A and 2 B assignments were made, any permutation of remaining 4 A assignments and 7 B assignments is equally likely. Thus, $v_1(3, 2) = 4/11$, $v_2(3, 2) = 7/11$. These conditional probabilities differ from the unconditional probabilities $w_1 = 7/16$, $w_2 = 9/16$.

Suppose the clinical study uses the allocation sequence $\omega = (\omega_1, ..., \omega_N)$. Let us denote by $v_{1l}(X_{l-1}, Y_{l-1})$ and $v_{2l}(X_{l-1}, Y_{l-1})$ the conditional probabilities of A and B allocations, respectively, at the lth allocation given the treatment totals X_{l-1} and Y_{l-1} in treatment groups A and B after the previous $l-1$ assignments. When making a guess regarding the ith allocation, the investigator who knows X_{i-1} and Y_{i-1} also knows $v_{1l}(X_{i-1}, Y_{i-1})$ and $v_{2l}(X_{i-1}, Y_{i-1})$.

The investigator wants to make a guess g_i that maximizes $D_1(g,\omega)$ (defined by Equation 8.3) given (X_{i-1}, Y_{i-1}).

The expected selection bias $D_1(g,\omega)$ given (X_{i-1}, Y_{i-1}) can be written as

$$
E(D_1(g,\omega) \mid X_{i-1}, Y_{i-1}) = \frac{2\Delta}{N/2} E\left(\frac{N}{2} (\alpha(g,\omega) + \beta(g,\omega) - 1) \mid X_{i-1}, Y_{i-1} \right) = \frac{2\Delta}{N/2}
$$

$$
E\left(\left(\sum_{l=1}^{i-1} \frac{1}{2} \left(\frac{N}{N_1} g_l \omega_l + \frac{N}{N_2} (1 - g_l)(1 - \omega_l) - 1 \right) \right) + \frac{1}{2} \left(\frac{N}{N_1} g_i \omega_i + \frac{N}{N_2} (1 - g_i)(1 - \omega_i) - 1 \right) \right.
$$

$$
\left. + \sum_{l=i+1}^{N} \frac{1}{2} \left(\frac{N}{N_1} g_l \omega_l + \frac{N}{N_2} (1 - g_l)(1 - \omega_l) - 1 \right) \mid X_{i-1}, Y_{i-1} \right).
$$

$$(8.16)$$

The only term in the sum above that depends on guess g_i is

$$
\frac{2\Delta}{N/2} E\left(\frac{1}{2} \left(\frac{N}{N_1} g_i \omega_i + \frac{N}{N_2} (1 - g_i)(1 - \omega_i) - 1 \right) \mid X_{i-1}, Y_{i-1} \right).
$$

We will call

$$
F_{Ei}^1(X_{i-1}, Y_{i-1}) = E\left(\frac{1}{2} \left(\frac{N}{N_1} g_i \omega_i + \frac{N}{N_2} (1 - g_i)(1 - \omega_i) - 1 \right) \mid X_{i-1}, Y_{i-1} \right) \quad (8.17)
$$

the Type 1 increment in the exact expected bias factor due to the ith allocation given (X_{i-1}, Y_{i-1}).

To maximize the selection bias $D_1(g,\omega)$ given (X_{i-1}, Y_{i-1}), the investigator needs to maximize $F_{Ei}^1(X_{i-1}, Y_{i-1})$, that is to maximize $t_{1i} + t_{2i}$, where

$$
t_{1i} = E\left(\frac{1}{N_1} g_i \omega_i \mid X_{i-1}, Y_{i-1} \right) = g_i P\{\omega_i = 1\} \sum_{j=1+X_{i-1}}^{N-(X_{i-1}+Y_{i-1})} \frac{1}{j} P\{N_1 = j \mid \omega_i = 1\}
$$

$$
= g_i v_{1i}(X_{i-1}, Y_{i-1}) \sum_{j=1+X_{i-1}}^{N-(X_{i-1}+Y_{i-1})} \frac{1}{j} P\{N_1 = j \mid \omega_i = 1\}
$$

$$(8.18)$$

and

$$t_{2i} = E\left(\frac{1}{N_2}(1 - g_i)(1 - \omega_i)|X_{i-1}, Y_{i-1}\right) = P\{\omega_i = 0\}$$

$$(1 - g_i) \sum_{j=Y_{i-1}+1}^{N-(X_{i-1}+Y_{i-1})} \frac{1}{j} P\{N_2 = j|\omega_i = 0\} =$$

$$= (1 - g_i)v_{2i}(X_{i-1}, Y_{i-1}) \sum_{Y_{i-1}+1}^{N-(X_{i-1}+Y_{i-1})} \frac{1}{j} P\{N_2 = j|\omega_i = 0\}.$$

(8.19)

From Equations 8.18 and 8.19, to maximize $t_{1i} + t_{2i}$, and thus $F_{Ei}^1(X_{i-1}, Y_{i-1})$ (Equation 8.17), the following choice of g_i should be made:

a. $g_i = 1$, if

$$v_{1i}(X_{i-1}, Y_{i-1}) \sum_{j=1+X_{i-1}}^{N-(X_{i-1}+Y_{i-1})} \frac{1}{j} P\{N_1 = j|\omega_i = 1\} > v_{2i}(X_{i-1}, Y_{i-1}) \sum_{Y_{i-1}+1}^{N-(X_{i-1}+Y_{i-1})} \frac{1}{j} P\{N_2 = j|\omega_i = 0\}$$

(8.20)

b. $g_i = 0$, if

$$v_{1i}(X_{i-1}, Y_{i-1}) \sum_{j=1+X_{i-1}}^{N-(X_{i-1}+Y_{i-1})} \frac{1}{j} P\{N_1 = j|\omega_i = 1\} < v_{2i}(X_{i-1}, Y_{i-1}) \sum_{Y_{i-1}+1}^{N-(X_{i-1}+Y_{i-1})} \frac{1}{j} P\{N_2 = j|\omega_i = 0\}.$$

(8.21)

c. Any $g_i = [0,1]$ will result in the same $t_{1i} + t_{2i}$ if

$$v_{1i}(X_{i-1}, Y_{i-1}) \sum_{j=1+X_{i-1}}^{N-(X_{i-1}+Y_{i-1})} \frac{1}{j} P\{N_1 = j|\omega_i = 1\} = v_{2i}(X_{i-1}, Y_{i-1}) \sum_{Y_{i-1}+1}^{N-(X_{i-1}+Y_{i-1})} \frac{1}{j} P\{N_2 = j|\omega_i = 0\}.$$

(8.22)

We will call the strategy defined by Rules 8.20 through 8.22 the exact conditional strategy.

8.3.2 Justification of the Directional Guessing Strategy

Equations 8.18 and 8.19 and Rules 8.20 through 8.22 are much simplified if the assumption $N_1 = \omega_1 N$ (which implies $N_2 = \omega_2 N$) is made (similar to how it is made for the equal allocation case [1]).

In this case, the Type 1 increment in the expected bias factor (no longer exact) given (X_{l-1}, Y_{l-1}) is

$$F_i^1(X_{i-1}, Y_{i-1}) = E\left[\frac{1}{2}\left(\frac{g_i}{w_1}\omega_i + \frac{(1-g_i)}{w_2}(1-\omega_i) - 1\right)\Big| X_{i-1}, Y_{i-1}\right]. \qquad (8.23)$$

Also,

$$t_{1i} = g_i v_{1i}(X_{i-1}, Y_{i-1})\frac{1}{w_1 N} \qquad (8.24)$$

$$t_{2i} = (1 - g_i)v_{2i}(X_{i-1}, Y_{i-1})\frac{1}{w_2 N} \qquad (8.25)$$

and thus, from Equations 8.24 and 8.25, $t_{1i} + t_{2i}$ and, therefore, $F_i^1(X_{i-1}, Y_{i-1})$ (Equation 8.23) is maximized by the following choice of g_i:

a. $g_i = 1$, if

$$\frac{v_{1i}(X_{i-1}, Y_{i-1})}{w_1} > \frac{v_{2i}(X_{i-1}, Y_{i-1})}{w_2}. \qquad (8.26)$$

b. $g_i = 0$, if

$$\frac{v_{1i}(X_{i-1}, Y_{i-1})}{w_1} < \frac{v_{2i}(X_{i-1}, Y_{i-1})}{w_2}. \qquad (8.27)$$

c. Any $g_i = [0,1]$ will result in the same $t_{1i} + t_{2i}$ if

$$\frac{v_{1i}(X_{i-1}, Y_{i-1})}{w_1} = \frac{v_{2i}(X_{i-1}, Y_{i-1})}{w_2}. \qquad (8.28)$$

The choice of g_i (Equations 8.26 through 8.28) constitutes the directional strategy where the treatment is guessed when the conditional probability of its allocation exceeds the unconditional probability [2]. When the conditional probabilities are equal to the unconditional probabilities (as is the case with the first allocation for any ARP procedure), similar to convergence strategy, the choice of g_i is made at random in $v_{1i}(X_{i-1}, Y_{i-1}):v_{2i}(X_{i-1}, Y_{i-1})$ ratio. However, the strategy where treatment A is chosen in such cases will lead to the same selection bias and can be used to simplify the calculations of the selection bias.

With the directional strategy, the Type 1 increment in the expected bias factor due to the ith allocation given (X_{i-1}, Y_{i-1}) is

$$F_i^1(X_{i-1}, Y_{i-1}) = \frac{1}{2}\left(\max\left(\frac{v_{1i(X_{i-1}, Y_{i-1})}}{w_1} , \frac{v_{1i(X_{i-1}, Y_{i-1})}}{w_2} \right) - 1 \right). \tag{8.29}$$

It should be noted that the equal split of 1 in the top part of Equation 8.16 among the N terms is completely arbitrary. However, such split leads to a nice interpretation of the Type 1 increment in the expected bias due to the ith allocation given (X_{i-1}, Y_{i-1}): when the choice of g_i cannot influence $F_i^1(X_{i-1}, Y_{i-1})$ (as in Equation 8.28), the Type 1 increment in the exact expected bias is 0. This means that no selection bias can be introduced with the ith allocation.

The justification of the directional guessing strategy can also be provided using Equation 8.12 for $D_1(g,\omega)$ based on the proportions of correct and incorrect guesses in groups A and B. In this case and under the assumption $N_1 = \omega_1 N$,

$$E(D_1(g,\omega)|X_{i-1}, Y_{i-1}) = \frac{2\Delta}{N/2}$$

$$E\left(\frac{N}{4}(\alpha(g,\omega) - \alpha'(g,\omega) + \beta(g,\omega) - \beta'(g,\omega))|X_{i-1}, Y_{i-1} \right)$$

$$= \frac{2}{N/2} \sum_{l=1}^{N} E\left(\frac{1}{4}\left(\frac{1}{w_1}(2g_i - 1)\omega_i + \frac{1}{w_2}(1 - 2g_i)(1 - \omega_i) \right)|X_{i-1}, Y_{i-1} \right).$$

Only the ith term in the sum above depends on the guess g_i. We will call it the Type 2 increment in the expected bias factor due to the ith allocation given (X_{i-1}, Y_{i-1}):

$$F_i^2(X_{i-1}, Y_{i-1}) = E\left(\frac{1}{4}\left(\frac{1}{w_1}(2g_i - 1)\omega_i + \frac{1}{w_2}(1 - 2g_i)(1 - \omega_i) \right)|X_{i-1}, Y_{i-1} \right). \tag{8.30}$$

It is easy to show that $F_i^2(X_{i-1}, Y_{i-1})$ (Equation 8.30) is also maximized under Equations 8.26 through 8.28, that is, under the directional strategy, and that in this case it takes on value

$$F_i^2(X_{i-1}, Y_{i-1}) = \frac{1}{4} \left| \frac{v_{1i}\ (X_{i-1}, Y_{i-1})}{w_1} - \frac{v_{2i}\ (X_{i-1}, Y_{i-1})}{w_2} \right|, \tag{8.31}$$

while for equal allocation,

$$F_i^1(X_{i-1}, Y_{i-1}) = F_i^2(X_{i-1}, Y_{i-1}) = \left| v_{1i}\ (X_{i-1}, Y_{i-1}) - \frac{1}{2} \right|, \tag{8.32}$$

and for unequal allocation,

$$F_i^1(X_{i-1}, Y_{i-1}) \neq F_i^2(X_{i-1}, Y_{i-1}).$$

It can be demonstrated, however, that for an unequal allocation ARP procedure,

$$EF_i^1(X_{i-1}, Y_{i-1}) = EF_i^2(X_{i-1}, Y_{i-1}) \text{ for all } i = 1, \ldots, N. \tag{8.33}$$

Both the Type 1 and Type 2 increments in the expected bias factor due to the ith allocation given (X_{i-1}, Y_{i-1}) can be used by the investigator to evaluate the payoff on the effort of biasing the ith allocation.

To illustrate how the directional strategy works for the PBR, we will visualize an allocation sequence as a path along the integer grid in the two-dimensional Cartesian plane as described in Ref. [29]. The horizontal axis represents allocation to treatment A; the vertical axis represents allocation to treatment B. The allocation path starts at the origin (0, 0) and with each allocation moves one unit along the axis that corresponds to the assigned treatment. After i allocations, the allocation path ends up at the node with coordinates (X_i, Y_i), where X_i and Y_i are the numbers of A and B allocations, respectively, within the first i allocations.

For the allocation procedure with the $C_1 : C_2$ allocation ratio, we will call the ray AR = $(C_1 u, C_2 u)$, $u \geq 0$, the allocation ray. For equal allocation procedures, the allocation ray is the diagonal of the non-negative quadrant.

Figure 8.1 presents the example of the PBR with a 7:9 allocation ratio and a PBS of 16. The allocation space that corresponds to the first block of 16 allocations is depicted by the black rectangle with the allocation ray cutting through its diagonal. For the PBR nodes above the allocation ray, the conditional probability of A allocation exceeds its unconditional probability, and therefore, treatment A is guessed for the next allocation. For the nodes

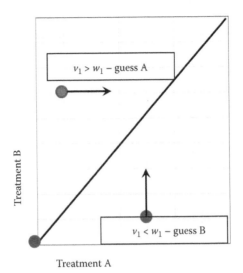

Treatment A

FIGURE 8.1
Directional strategy with 7:9 PBR allocation. The solid rectangle denotes the allocation space for the first PBR block; the diagonal black line denotes the allocation ray.

below the allocation ray, the conditional probability of A allocation is below its unconditional probability, and therefore, treatment B is guessed for the next allocation. For the nodes on the allocation ray (at the beginning of a permuted block), the random guess is made in a 7:9 ratio.

For the equal allocation PBR, the directional strategy reverts to the convergence strategy where the treatment with the lowest total is guessed. This is also the case for other equal allocation procedures for which $v_{1i}(X_{i-1}, Y_{i-1}) > v_{2i}(X_{i-1}, Y_{i-1})$ at the nodes above the allocation ray and $v_{1i}(X_{i-1}, Y_{i-1}) < v_{2i}(X_{i-1}, Y_{i-1})$ below the allocation ray, such as biased coin randomization [30], biased coin randomization with imbalance tolerance [31], accelerated biased coin design [32], Wei's urn design [33,34], block urn design [27], and others. For other allocation procedures, for example, the truncated binomial design [1] or the big stick design [5], $v_{1i}(X_{i-1}, Y_{i-1}) = v_{2i}(X_{i-1}, Y_{i-1})$, there are areas above and below the allocation ray where the convergence strategy guess is the underrepresented treatment, while the directional strategy guess is made by a toss of a fair coin. However, with these allocation procedures, the convergence strategy will result in the same selection bias as the directional strategy.

As we demonstrated, the justification of the directional strategy requires the assumption $N_1 = \omega_1 N$, $N_2 = \omega_2 N$. When this assumption does not hold, the directional strategy most often, but not always, agrees with the exact conditional strategy. When the directional strategy disagrees with the exact conditional strategy, the directional strategy does not lead to the highest selection bias unless any choice of $g_i = [0,1]$ provides the same selection bias with the exact conditional strategy (Rule 8.22).

A simple example of such disagreement is provided by the study with 3:10 PBR ($w_1 = 3/13$, $w_2 = 10/13$) where a certain site is supposed to allocate 10 subjects. Suppose of the first nine subjects at the site, two subjects are assigned to group A and seven subjects are assigned to group B. Then, $v_{1,10}(2,7) = 1/4$ and $v_{2,10}(2,7) = 3/4$. Conditional on the 10th allocation being A allocation ($\omega_{10} = 1$), $N_1 = 3$ with probability 1; conditional on the 10th allocation being B allocation ($\omega_{10} = 0$), $N_2 = 8$ with probability 1.

Since $v_{1,10}(2,7)/w_1 = 1/4:3/13 > 1$, the directional strategy dictates the 10th allocation guess to be A. However,

$$v_{1,10}(2,7)\sum_{j=3}^{3}\frac{1}{j}P\{N_1 = j|\omega_{10} = 1\} = 1/4 \times 1/3 = 1/12,$$

$$v_{2,10}(2,7)\sum_{j=8}^{8}\frac{1}{j}P\{N_2 = j|\omega_{10} = 0\} = 3/4 \times 1/8 = 3/32,$$

and since

$$v_{1,10}(2,7)\sum_{j=3}^{3}\frac{1}{j}P\{N_1 = j|\omega_{10} = 1\} < v_{2,10}(2,7)\sum_{j=8}^{8}\frac{1}{j}P\{N_2 = j|\omega_{10} = 0\},$$

per Rule 8.21 the exact conditional strategy dictates the 10th allocation guess to be B.

Although for small N the directional strategy might disagree with the exact conditional strategy and thus fail to maximize the selection bias, it can be justified on practical grounds. The exact conditional rules are too cumbersome for the investigator to calculate and thus can hardly be expected to be used in practice. Also, the exact number of subjects that an investigator will enroll is often unknown and the exact conditional strategy cannot be employed. This is the case in the multi-center trials with competitive enrollment or a single-center trial stratified by baseline factors, where the size of an individual stratum can vary.

Thus, the directional strategy is often a sensible choice for large as well as small N. The loss in the expected selection bias due to rare instances of the discrepancies between the directional and exact conditional strategies is likely to be small. We will not pursue further exploration of this topic in this chapter.

8.3.3 Expected Bias Factor under the Directional Strategy

Rosenberger and Lachin [1] link the expected bias factor $E(F)$ under the convergence strategy for an equal allocation procedure to measures of predictability of the allocation procedure used by other authors [29,35]. Following Rosenberger and Lachin [1, p. 83], we will provide similar links for unequal allocation procedures under the directional strategy.

Under the assumption $N_1 = \omega_1 N$, $N_2 = \omega_2 N$, from Equations 8.9, 8.1, and 8.2,

$$E(F) = \frac{N}{2}(\alpha + \beta - 1) = \frac{N}{2}\left(E\left[\sum_{i=1}^{N} \frac{1}{N_1} g_i \omega_i + \frac{1}{N_2}(1 - g_i)(1 - \omega_i) \right] - 1 \right)$$

$$= \frac{1}{2} E\left[\sum_{i=1}^{N} \frac{1}{w_1} g_i \omega_i + \frac{1}{w_2}(1 - g_i)(1 - \omega_i) \right] - \frac{N}{2} = \frac{1}{2} EE$$

$$\left(\sum_{i=1}^{N} \frac{1}{w_1} g_i \omega_i + \frac{1}{w_2}(1 - g_i)(1 - \omega_i) \middle| v_{11}(0,0), \ldots, v_{1N}(X_{N-1}, Y_{N-1}) \right)$$

$$-\frac{N}{2} = \frac{1}{2} E \sum_{i=1}^{N} \left(\frac{1}{w_1} I(v_{1i}(X_{i-1}, Y_{i-1}) > w_1) + \frac{1}{w_2} I(v_{1i}(X_{i-1}, Y_{i-1}) < w_1) \right.$$

$$+ I(v_{1i}(X_{i-1}, Y_{i-1}) = w_1)\left(\frac{1}{w_1} w_1 w_1 + \frac{1}{w_2} w_2 w_2 \right) \right) - \frac{N}{2}.$$

Thus, under the directional strategy,

$$E(F) = \frac{1}{2} \sum_{i=1}^{N} E\left(\max\left(\frac{v_{1i}(X_{i-1}, Y_{i-1})}{w_1}, \frac{v_{2i}(X_{i-1}, Y_{i-1})}{w_2} \right) \right) - \frac{N}{2}. \qquad (8.34)$$

The expected bias factor is the sum of the expectations of the Type 1 increments in the expected bias factor (Equation 8.29) across the N generations:

$$E(F) = \sum_{i=1}^{N} E\left(F_i^1(X_{i-1}, Y_{i-1}) \right). \qquad (8.35)$$

For equal allocation where $E(F)$ is expressed as Equation 8.34, $E(F)/N$ reverts to

$$E(F)/N = \frac{1}{N} \sum_{i=1}^{N} E(\max(v_{1i}(X_{i-1}, Y_{i-1}), v_{2i}(X_{i-1}, Y_{i-1}))) - \frac{1}{2}. \qquad (8.36)$$

Chen (Equation 8.35) calls the right part of Equation 8.36 the average excess selection bias (AESB). The concept of the AESB is useful when comparing selection bias across allocation procedures based on blocks of different size, in particular, unequal allocation PBR with various allocation ratios. We will expand the definition of the AESB to unequal allocation procedures as

$$\gamma = \frac{E(F)}{N} = \frac{1}{2N} \sum_{i=1}^{N} E\left(\max\left(\frac{v_{1i}\,(X_{i-1},Y_{i-1})}{w_1}, \frac{v_{1i}\,(X_{i-1},Y_{i-1})}{w_2} \right) \right) - \frac{1}{2}. \quad (8.37)$$

The expected bias factor under the directional strategy can also be expressed through the proportions of correct and incorrect guesses in the two treatment groups. From Equations 8.15, 8.1, 8.2, 8.10, and 8.11,

$$E(F) = \frac{N}{4}(\alpha - \alpha' + \beta - \beta')$$

$$= \frac{N}{4}E\left(\sum_{i=1}^{N} \frac{1}{N_1}(g_i - (1-g_i))\omega_i + \frac{1}{N_2}((1-g_i) - g_i)(1-\omega_i) \right)$$

$$= \frac{1}{4}EE\left(\sum_{i=1}^{N} \left(\frac{1}{w_1}(2g_i - 1) \right)\omega_i + \frac{1}{w_2}(1 - 2g_i)(1-\omega_i) \right) | v_{11}(0,0), \dots, v_{1N}(X_{N-1},Y_{N-1}) \right)$$

$$= \frac{1}{4}E\sum_{i=1}^{N} \left(I(v_{1i}(X_{i-1},Y_{i-1}) > w_1)\left(\frac{v_{1i}(X_{i-1},Y_{i-1})}{w_1} - \frac{v_{2i}(X_{i-1},Y_{i-1})}{w_2} \right) \right.$$

$$+ I(v_{1i}(X_{i-1},Y_{i-1}) < w_1)\left(-\frac{v_{1i}(X_{i-1},Y_{i-1})}{w_1} + \frac{v_{2i}(X_{i-1},Y_{i-1})}{w_2} \right)$$

$$+ I(v_{1i}(X_{i-1},Y_{i-1}) = w_1)\left(\frac{1}{w_1}v_{1i}(X_{i-1},Y_{i-1})(v_{1i}(X_{i-1},Y_{i-1}) - v_{2i}(X_{i-1},Y_{i-1})) \right.$$

$$\left. + \frac{1}{w_2}v_{2i}(X_{i-1},Y_{i-1})(v_{2i}(X_{i-1},Y_{i-1}) - v_{1i}(X_{i-1},Y_{i-1})) \right) \right).$$

Thus,

$$E(F) = \frac{1}{4} \sum_{i=1}^{N} E\left| \frac{v_{1i}\,(X_{i-1},Y_{i-1})}{w_1} - \frac{v_{2i}\,(X_{i-1},Y_{i-1})}{w_2} \right| \quad (8.38)$$

or

$$E(F) = \sum_{i=1}^{N} E\Big(F_i^2(X_{i-1}, Y_{i-1})\Big). \tag{8.39}$$

Thus, γ can also be expressed as

$$\gamma = \frac{E(F)}{N} = \frac{1}{4N} \sum_{i=1}^{N} E\left| \frac{v_{1i}(X_{i-1}, Y_{i-1})}{w_1} - \frac{v_{2i}(X_{i-1}, Y_{i-1})}{w_2} \right|. \tag{8.40}$$

For equal allocation, the right side of Equation 8.38 reverts to ρ_{pred} from Equation 5.5 in Ref. [1]:

$$\rho_{pred} = \sum_{i=1}^{N} E\left| v_{1i}(X_{i-1}, Y_{i-1}) - \frac{1}{2} \right|. \tag{8.41}$$

Berger et al. [29] use Equation 8.41 to measure the predictability of an equal allocation restricted randomization procedure. Rosenberger and Lachin (2016) show that it is equivalent to the expected bias factor for equal allocation; we followed their steps to prove Equations 8.34 and 8.38 for unequal allocation.

8.4 Selection Bias with PBR

In this section, we will calculate the AESB for the two-arm PBR under the directional guessing strategy and discuss its properties. For equal allocation PBR, Rosenberger and Lachin (2016) provide an explicit formula for the expected bias factor (Equation 5.6). From it and Equation 8.40, the AESB in a study with $n/2$ subjects allocated to group A and $n/2$ subjects allocated to group B is

$$\gamma = \frac{1}{n} \left(\frac{2^{n-1}}{\binom{n}{n/2}} - \frac{1}{2} \right), \tag{8.42}$$

where $\binom{i}{j}$ is the number of j combinations from a set of i items.

For unequal allocation, the explicit formula for the selection bias factor under the directional strategy is not derived (possibly, due to a lack of interest).

We will call the probability for an allocation sequence to reside in the node (X, Y) after i allocations the resident probability of the node (X, Y) and denote it $R(X, Y)$.

For the two-arm PBR with $C_1{:}C_2$ allocation ratio (or with allocation probabilities w_1 and w_2), let us denote by $Q = mS$ the PBS, and by $Q_1 = mC_1$, $Q_2 = mC_2$ the dimensions of the permuted block ($Q_1 + Q_2 = Q$). Then, the resident probability of the node $R(X,Y)$ is

$$R(X,Y) = \frac{\binom{X+Y}{X}\binom{Q-(X+Y)}{Q_1-X}}{\binom{Q}{Q_1}}. \tag{8.43}$$

The resident probabilities in the example of 7:9 PBR with a PBS of 16 are tabulated in Table 8.1 for the nodes (X, Y) within the first block. The rows that represent Y from 0 to 9 are placed in decreasing order to mimic the placement of the nodes on the two-dimensional allocation plane. As one can see, the resident probabilities within generation $j = X + Y$, $j = 0,...$, 16 are high around the allocation ray (the diagonal of the block) and decrease in the direction away from the allocation ray.

For the PBR, the conditional probabilities of treatment A and B allocation are $v_1 = (Q_1 - X)/(Q - (X + Y))$ and $v_2 = (Q_2 - Y)/(Q - (X + Y))$, respectively.

TABLE 8.1

The Resident Probabilities for 7:9 PBR with a PBS of 16

9	0.0001	0.0009	0.0048	0.0192	0.0625	0.1750	0.4375	1.0000
8	0.0007	0.0055	0.0236	0.0721	0.1731	0.3375	0.5250	0.5625
7	0.0031	0.0196	0.0661	0.1573	0.2885	0.4154	0.4500	0.3000
6	0.0105	0.0514	0.1371	0.2570	0.3671	0.4038	0.3231	0.1500
5	0.0288	0.1101	0.2313	0.3427	0.3855	0.3304	0.2019	0.0692
4	0.0692	0.2019	0.3304	0.3855	0.3427	0.2313	0.1101	0.0288
3	0.1500	0.3231	0.4038	0.3671	0.2570	0.1371	0.0514	0.0105
2	0.3000	0.4500	0.4154	0.2885	0.1573	0.0661	0.0196	0.0031
1	0.5625	0.5250	0.3375	0.1731	0.0721	0.0236	0.0055	0.0007
0	1.0000	0.4375	0.1750	0.0625	0.0192	0.0048	0.0009	0.0001
Y/X	0	1	2	3	4	5	6	7

Thus, the Type 1 increment in the expected bias factor (Equation 8.29) after X allocations to A and Y allocations to B $(X \leq Q_1, Y \leq Q_2)$ is

$$F_i^1(X,Y) = \frac{1}{2}\left(\max\left(\frac{Q_1 - X}{(Q-(X+Y))w_1}, \frac{Q_2 - Y}{(Q-(X+Y))w_2}\right) - 1\right)$$

and the Type 2 increment in the expected bias factor (Equation 8.31) is

$$F_i^2(X,Y) = \frac{1}{4}\left|\frac{Q_1 - X}{(Q-(X+Y))w_1} - \frac{Q_2 - Y}{(Q-(X+Y))w_2}\right|. \tag{8.44}$$

The Type 2 increments in the expected bias factor (Equation 8.44) for the example of 7:9 PBR with a PBS of 16 are tabulated in Table 8.2 for the nodes (X, Y) within the first block. The Type 1 and Type 2 increments are very close for all nodes (X, Y) as is the case for all allocation ratios close to 1:1. Within a generation, the increments are low near the allocation ray and increase in the directions away from it. To illustrate this point, the cells around the diagonal that have the Type 2 increment below 0.2 are highlighted.

This shows that the investigator who tries to bias the allocation from a node close to the allocation ray gets a very small payoff on his effort in the form of the increment in the expected bias factor, while biasing from a node far removed from the allocation ray brings a considerable payoff. These considerations lead to a strategy where the investigator biases the allocation only when the expected payoff is large (to be discussed in Section 8.5).

For equal allocation, the maximum Type 2 increment in the expected bias factor is 0.5. It is reached when the allocation to one of the treatments is

TABLE 8.2

The Type 2 Increment in the Expected Bias Factor with the Directional Guessing Strategy for 7:9 PBR with the PBS = 16 as a Function of Achieved Treatment Totals

9	0.5714	0.5714	0.5714	0.5714	0.5714	0.5714	0.5714	0.0000
8	0.4444	0.4263	0.4021	0.3683	0.3175	0.2328	0.0635	0.4444
7	0.3457	0.3175	0.2812	0.2328	0.1651	0.0635	0.1058	0.4444
6	0.2667	0.2328	0.1905	0.1361	0.0635	0.0381	0.1905	0.4444
5	0.2020	0.1651	0.1199	0.0635	0.0091	0.1058	0.2413	0.4444
4	0.1481	0.1097	0.0635	0.0071	0.0635	0.1542	0.2751	0.4444
3	0.1026	0.0635	0.0173	0.0381	0.1058	0.1905	0.2993	0.4444
2	0.0635	0.0244	0.0212	0.0750	0.1397	0.2187	0.3175	0.4444
1	0.0296	0.0091	0.0537	0.1058	0.1674	0.2413	0.3316	0.4444
0	0.0000	0.0381	0.0816	0.1319	0.1905	0.2597	0.3429	0.4444
Y/X	0	1	2	3	4	5	6	7

deterministic. For unequal allocation, however, a deterministic allocation to treatment A corresponds to the Type 2 increment in the expected bias factor of $1/(4w_1)$, while a deterministic allocation to treatment B corresponds to the Type 2 increment of $1/(4w_2)$. Thus, while for equal allocation PBR the Type 2 increments in the expected bias factor on the top and right borders of the permuted block are all equal to 0.5, for the unequal allocation PBR, the Type 2 increments on the top border are $1/(4w_1)$ and those on the right border are $1/(4w_2)$. For the $w_j = \min(w_1, w_2)$, the increment $1/(4w_j) > 0.5$, while for the $w_k = \max(w_1, w_2)$, the increment $1/(4w_k) < 0.5$. The maximum Type 2 increment in the expected bias factor is thus $1/(4\min(w_1, w_2))$.

From Equations 8.35 and 8.39,

$$E(F) = \sum_{i=1}^{N} E\left(F_i^1(X,Y)\right) = \sum_{i=1}^{N} E\left(F_i^2(X,Y)\right) = \sum_{\substack{X=0 \text{ to } Q_1 \\ Y=0 \text{ to } Q_2 \\ X+Y<Q}} R(X,Y)F_i^2(X,Y),$$

where $R(X,Y)$ and $F_i^2(X,Y)$ are defined by Equations 8.43 and 8.44, respectively.

Figure 8.2 presents the expected increment to the expected bias factor $E\left(F_i^2(X,Y)\right)$ by generation—which is the same regardless of whether the Type 1 or Type 2 increment to the expected bias factor is used according to Equation 8.33. We see that the contribution of the generations to the expected bias factor is monotonically non-decreasing across the generations, with the

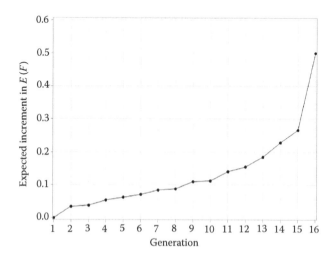

FIGURE 8.2
Expected increment in the expected bias factor $E\left(F_i^2(X,Y)\right)$ for 7:9 PBR with the PBS = 16 by generation.

generation before the end of the block (generation 15) contributing 0.5 to the expected bias factor.

From Equation 8.37, the AESB across the first Q allocations is

$$\gamma_{PBR}(Q_1, Q_2) = \frac{1}{2N} \sum_{\substack{X=0 \text{ to } Q_1 \\ Y=0 \text{ to } Q_2 \\ X+Y<Q}} R(X,Y) \max\left(\frac{v_1(X,Y)}{w_1}, \frac{v_2(X,Y)}{w_2}\right) - \frac{1}{2}$$

$$= \frac{1}{2N} \sum_{\substack{X=0 \text{ to } Q_1 \\ Y=0 \text{ to } Q_2 \\ X+Y<Q}} \frac{\binom{X+Y}{X}\binom{Q-(X+Y)}{Q_1-X}}{\binom{Q}{Q_1}} \max\left(\frac{v_1(X,Y)}{w_1}, \frac{v_2(X,Y)}{w_2}\right) - \frac{1}{2}.$$

Alternatively, from Equation 8.40, AESB can be expressed as

$$\gamma_{PBR}(Q_1, Q_2) = \frac{1}{2N} \sum_{\substack{X=0 \text{ to } Q_1 \\ Y=0 \text{ to } Q_2 \\ X+Y<Q}} R(X,Y)\left|\frac{v_1(X,Y)}{w_1} - \frac{v_2(X,Y)}{w_2}\right|$$

$$= \frac{1}{2N} \sum_{\substack{X=0 \text{ to } Q_1 \\ Y=0 \text{ to } Q_2 \\ X+Y<Q}} \frac{\binom{X+Y}{X}\binom{Q-(X+Y)}{Q_1-X}}{\binom{Q}{Q_1}}\left|\frac{v_1(X,Y)}{w_1} - \frac{v_2(X,Y)}{w_2}\right|.$$

(8.45)

The AESB calculated according to Equation 8.45 is tabulated for Q_1 and Q_2 from 1 to 10 in Table 8.3. It is symmetric with respect to Q_1 and Q_2. The diagonal cells in Table 8.3 represent equal allocation and thus match Equation 8.42. For equal allocation, the AESB is decreasing as the PBS increases (in accordance with Equation 8.42). Pairs (Q_1, Q_2) with the same PBS = Q are located along the straight line connecting $(1, Q - 1)$ and $(Q - 1, 1)$. It is easy to see that among all pairs (Q_1, Q_2) with the same PBS = Q, the AESB is the smallest for equal allocation and grows as the allocation becomes more and more unbalanced. For the most unbalanced allocation of $(1:Q - 1)$ and $(Q - 1:1)$, the AESB is 0.25.

It is interesting to note that the AESB for $(m + 1):m$ and for $m:(m + 1)$ allocation is the same as that for $m:m$ allocation.

Let us take a note of the AESB for the 1:1 PBR with a PBS of 4—the most commonly used allocation procedure. The 1:1 PBR with a PBS of 2 is rarely used in open-label trails because of its high predictability. However, the AESB

TABLE 8.3

Average Excess Selection Bias under the Directional Strategy for $Q_1:Q_2$ PBR with Q_1 and Q_2 from 1 to 10

Q_1/Q_2	1	2	3	4	5	6	7	8	9	10
1	0.2500	0.2500	0.2500	0.2500	0.2500	0.2500	0.2500	0.2500	0.2500	0.2500
2	0.2500	0.2083	0.2083	0.2000	0.2000	0.1964	0.1964	0.1944	0.1944	0.1932
3	0.2500	0.2083	0.1833	0.1833	0.1786	0.1726	0.1726	0.1708	0.1682	0.1682
4	0.2500	0.2000	0.1833	0.1661	0.1661	0.1617	0.1589	0.1545	0.1545	0.1530
5	0.2500	0.2000	0.1786	0.1661	0.1532	0.1532	0.1500	0.1472	0.1448	0.1414
6	0.2500	0.1964	0.1726	0.1617	0.1532	0.1430	0.1430	0.1402	0.1376	0.1358
7	0.2500	0.1964	0.1726	0.1589	0.1500	0.1430	0.1348	0.1348	0.1326	0.1304
8	0.2500	0.1944	0.1708	0.1545	0.1472	0.1402	0.1348	0.1279	0.1279	0.1259
9	0.2500	0.1944	0.1682	0.1545	0.1448	0.1376	0.1326	0.1279	0.1220	0.1220
10	0.2500	0.1932	0.1682	0.1530	0.1414	0.1358	0.1304	0.1259	0.1220	0.1169

of 0.2083 for the 1:1 PBR with a PBS of 4 is not much smaller than the AESB of the 1:1 PBR with a PBS of 2 (0.25). One has to use a PBS of 18 to cut the AESB of 0.25 at least in half within a PBR family.

Figure 8.3 illustrates the behavior of the AESB when Q_1 is held constant and Q_2 increases on the examples of Q_1 = 1, 2, 3, 4, 5, 25, 50, and 75, and Q_2 ranging from 1 to 90. As Q_2 increases for fixed Q_1, the AESB decreases and approaches a plateau.

In practice, very large block sizes are hardly ever used with the equal allocation as large imbalances can be observed. The large block sizes, however,

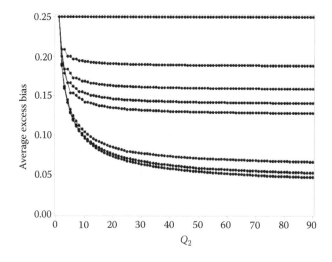

FIGURE 8.3

Average excess selection bias for $Q_1:Q_2$ PBR, for Q_1: 1, 2, 3, 4, 5, 25, 50, and 75 (lines from top to bottom).

can arise in studies with unequal allocation with the inconvenient allocation ratio. In this case, the PBR provides no better balance for small cohorts than the complete randomization [8,10]. The alternative is presented by the BTR [8,10]—an ARP procedure that ensures that the allocation sequences stay close to the allocation ray. Although for the investigator who is aware of the BTR being used in the study the potential for selection bias can be rather large, for the unaware investigator who expects the PBR and uses the directional strategy with a payoff threshold, the selection bias is low. This will be discussed in detail in Section 8.5.

8.5 Directional Strategy with the Payoff Threshold

As can be seen on the example of the 7:9 PBR, the increment in the expected bias factor with the directional guessing strategy is very low from the nodes near the allocation ray. At the same time, biasing is hard work: if the next available subject does not match the guessed group, they have to be sent home and not randomized into the study. Meanwhile, the investigator will wait until the proper match comes along. This does not promote a good collaboration with patients and slows down the enrollment. Thus, the investigator might choose to bias the randomization only when the payoff to the biasing effort, that is, the increment in the expected bias factor, is large.

This strategy is described for equal allocation in Refs. [2,4].

Suppose the investigator still selects a healthier patient with the expected response $\mu + \Delta$ when his guess for the next allocation is Treatment A and selects a sicker patient with the expected response $\mu - \Delta$ when his guess is Treatment B. However, when the payoff on the biasing effort is low, the investigator does not attempt to bias the allocation and allocates the next available patient with the expected response μ.

It is easy to show that, in this model, all the formulas that express the selection bias through the proportions of correct and incorrect guesses in the treatment groups A and B (specifically Equations 8.12, 8.13, and 8.14) as well as the exact expected selection bias factor formula (Equation 8.15) remain accurate.

Let $\chi_i = 1$ if a guess was made for the ith allocation; $\chi_i = 0$ otherwise.

Under the assumption $N_1 = \omega_1 N$, when biasing is applied, the Type 2 increment in the expected bias factor due to the ith allocation (Equation 8.30) is maximized by the directional guessing strategy (Equations 8.26 through 8.28). Under the directional strategy,

$$F_i^2(X_{i-1}, Y_{i-1}) = \frac{1}{4} \left| \frac{v_{1i}(X_{i-1}, Y_{i-1})}{w_1} - \frac{v_{2i}(X_{i-1}, Y_{i-1})}{w_2} \right| \chi_i. \tag{8.46}$$

The expected bias factor under the directional strategy is

$$E(F) = \sum_{i=1}^{N} \frac{1}{4} E \left| \frac{v_{1i}\,(X_{i-1},Y_{i-1})}{w_1} - \frac{v_{2i}\,(X_{i-1},Y_{i-1})}{w_2} \right| \chi_i. \tag{8.47}$$

The Type 2 increment in the expected bias factor due to the ith allocation (Equation 8.46) thus provides a good way to establish a biasing threshold T: the investigator biases the next allocation only if $F_i^2(X_{i-1}, Y_{i-1}) \geq T$.

For equal allocation, Berger [4] uses the biasing threshold $G > 1/2$ where the investigator biases only if $v_{1i}\,(X_{i-1},\,Y_{i-1}) \geq G$ or $v_{1i}\,(X_{i-1},\,Y_{i-1}) \leq 1 - G$.

Since for equal allocation $F_i^2(X_{i-1},Y_{i-1})$ reduces to $\left| v_{1i}\,(X_{i-1},Y_{i-1}) - \frac{1}{2} \right|$ per Equation 8.32, threshold G on the conditional probability of Treatment A allocation corresponds to the threshold $T = G - 1/2$ for $F_i^2(X_{i-1},Y_{i-1})$.

For equal allocation, Berger [4] uses $G = 0.99$ as an equivalent to deterministic allocation. For unequal allocation, a single threshold does not define the deterministic allocations. As we discussed earlier, when the allocation to A is deterministic, $F_i^2(X_{i-1},Y_{i-1}) = 1/(4w_1)$, while when the allocation to B is deterministic, $F_i^2(X_{i-1},Y_{i-1}) = 1/(4w_2)$.

From Equation 8.47, when the directional strategy with biasing threshold is used with the $Q_1{:}Q_2$ PBR, the AESB is

$$\gamma_{PBR}(Q_1,Q_2) = \frac{1}{2N} \sum_{\substack{X=0 \text{ to } Q_1 \\ Y=0 \text{ to } Q_2 \\ X+Y<Q}} R(X,Y) \left| \frac{v_1\,(X,Y)}{w_1} - \frac{v_2\,(X,Y)}{w_2} \right| I\left(\left| \frac{v_1(X,Y)}{w_1} - \frac{v_2(X,Y)}{w_2} \right| \geq T \right)$$

$$= \frac{1}{2N} \sum_{\substack{X=0 \text{ to } Q_1 \\ Y=0 \text{ to } Q_2 \\ X+Y<Q}} \frac{\binom{X+Y}{X}\binom{Q-(X+Y)}{Q_1-X}}{\binom{Q}{Q_1}} \left| \frac{v_1\,(X,Y)}{w_1} - \frac{v_2\,(X,Y)}{w_2} \right| I\left(\left| \frac{v_1\,(X,Y)}{w_1} - \frac{v_2\,(X,Y)}{w_2} \right| \geq T \right).$$

$$\tag{8.48}$$

Table 8.4 presents the AESB (Equation 8.48) under the directional strategy when the biasing threshold $T = 0.15$ is used for $Q_1{:}Q_2$ PBR with Q_1 and Q_2 from 1 to 10. This biasing threshold corresponds to $G = 0.65$ for equal allocation.

Table 8.5 presents the ratio of AESB with $T = 0.15$ over AESB without a threshold. The impact of the threshold is more pronounced for larger block sizes where more nodes around the allocation ray are removed from biasing

TABLE 8.4

Average Excess Selection Bias under the Directional Strategy with the Biasing Threshold of 0.15 for Q_1:Q_2 PBR with Q_1 and Q_2 from 1 to 10

Q_1/Q_2	1	2	3	4	5	6	7	8	9	10
1	0.2500	0.2500	0.2292	0.2375	0.2417	0.2321	0.2366	0.2292	0.2333	0.2364
2	0.2500	0.2083	0.1750	0.1833	0.1631	0.1622	0.1696	0.1625	0.1621	0.1674
3	0.2292	0.1750	0.1667	0.1506	0.1327	0.1488	0.1423	0.1379	0.1336	0.1265
4	0.2375	0.1833	0.1506	0.1464	0.1319	0.1258	0.1127	0.1260	0.1184	0.1138
5	0.2417	0.1631	0.1327	0.1319	0.1282	0.1136	0.1126	0.1026	0.1134	0.1102
6	0.2321	0.1622	0.1488	0.1258	0.1136	0.1158	0.1033	0.1003	0.1009	0.0912
7	0.2366	0.1696	0.1423	0.1127	0.1126	0.1033	0.1051	0.0949	0.0944	0.0907
8	0.2292	0.1625	0.1379	0.1260	0.1026	0.1003	0.0949	0.0968	0.0878	0.0875
9	0.2333	0.1621	0.1336	0.1184	0.1134	0.1009	0.0944	0.0878	0.0896	0.0818
10	0.2364	0.1674	0.1265	0.1138	0.1102	0.0912	0.0907	0.0875	0.0818	0.0835

TABLE 8.5

Ratio of the AESB under the Directional Strategy with Biasing Threshold 0.15 over the AESB under the Directional Strategy without a Threshold

Q_1/Q_2	1	2	3	4	5	6	7	8	9	10
1	1.0000	1.0000	0.9167	0.9500	0.9667	0.9286	0.9464	0.9167	0.9333	0.9455
2	1.0000	1.0000	0.8400	0.9167	0.8155	0.8258	0.8636	0.8357	0.8338	0.8667
3	0.9167	0.8400	0.9091	0.8214	0.7433	0.8621	0.8241	0.8075	0.7943	0.7519
4	0.9500	0.9167	0.8214	0.8817	0.7945	0.7779	0.7094	0.8154	0.7663	0.7442
5	0.9667	0.8155	0.7433	0.7945	0.8368	0.7415	0.7508	0.6968	0.7831	0.7795
6	0.9286	0.8258	0.8621	0.7779	0.7415	0.8096	0.7221	0.7152	0.7337	0.6718
7	0.9464	0.8636	0.8241	0.7094	0.7508	0.7221	0.7800	0.7038	0.7122	0.6956
8	0.9167	0.8357	0.8075	0.8154	0.6968	0.7152	0.7038	0.7567	0.6865	0.6953
9	0.9333	0.8338	0.7943	0.7663	0.7831	0.7337	0.7122	0.6865	0.7342	0.6703
10	0.9455	0.8667	0.7519	0.7442	0.7795	0.6718	0.6956	0.6953	0.6703	0.7146

when the threshold is applied. The AESB for the blocks (1:1), (1:2), (2:1), and (2:2) remains unaffected as for none of the nodes $F_i^2(X_{i-1}, Y_{i-1}) < 0.15$.

Tables 8.6 and 8.7 present the AESB under the directional strategy with the biasing threshold $T = 0.25$ (Table 8.5) and the corresponding ratio of AESB with $T = 0.25$ over the AESB without a threshold. The biasing threshold $T = 0.25$ corresponds to $G = 0.75$ for equal allocation. Again, we see a higher impact of the threshold for larger block sizes, with the ratio as low as 0.47 for the (10:10) permuted block. Across the permuted blocks with the same block size, the reduction in AESB caused by the threshold is the highest for equal allocation and diminishes as blocks become more and more unbalanced.

TABLE 8.6

Average Excess Selection Bias under the Directional Strategy with the Biasing
Threshold of 0.25 for $Q_1:Q_2$ PBR with Q_1 and Q_2 from 1 to 10

Q_1/Q_2	1	2	3	4	5	6	7	8	9	10
1	0.2500	0.2083	0.2292	0.2125	0.2250	0.2143	0.2232	0.2153	0.2222	0.2159
2	0.2083	0.1667	0.1625	0.1458	0.1631	0.1533	0.1533	0.1458	0.1561	0.1504
3	0.2292	0.1625	0.1250	0.1179	0.1182	0.1136	0.1062	0.1216	0.1159	0.1138
4	0.2125	0.1458	0.1179	0.1054	0.1002	0.1021	0.0988	0.0936	0.0882	0.1015
5	0.2250	0.1631	0.1182	0.1002	0.0909	0.0918	0.0923	0.0860	0.0836	0.0796
6	0.2143	0.1533	0.1136	0.1021	0.0918	0.0797	0.0818	0.0742	0.0765	0.0725
7	0.2232	0.1533	0.1062	0.0988	0.0923	0.0818	0.0713	0.0723	0.0674	0.0718
8	0.2153	0.1458	0.1216	0.0936	0.0860	0.0742	0.0723	0.0646	0.0658	0.0617
9	0.2222	0.1561	0.1159	0.0882	0.0836	0.0765	0.0674	0.0658	0.0591	0.0611
10	0.2159	0.1504	0.1138	0.1015	0.0803	0.0725	0.0718	0.0617	0.0611	0.0545

TABLE 8.7

Ratio of the AESB under the Directional Strategy with Biasing Threshold 0.25
over the AESB under the Directional Strategy without a Threshold

Q_1/Q_2	1	2	3	4	5	6	7	8	9	10
1	1.0000	0.8333	0.9167	0.8500	0.9000	0.8571	0.8929	0.8611	0.8889	0.8636
2	0.8333	0.8000	0.7800	0.7292	0.8155	0.7803	0.7803	0.7500	0.8026	0.7784
3	0.9167	0.7800	0.6818	0.6429	0.6617	0.6580	0.6149	0.7118	0.6892	0.6764
4	0.8500	0.7292	0.6429	0.6344	0.6033	0.6313	0.6219	0.6058	0.5710	0.6637
5	0.9000	0.8155	0.6617	0.6033	0.5933	0.5995	0.6156	0.5839	0.5772	0.5628
6	0.8571	0.7803	0.6580	0.6313	0.5995	0.5574	0.5719	0.5292	0.5561	0.5341
7	0.8929	0.7803	0.6149	0.6219	0.6156	0.5719	0.5293	0.5363	0.5081	0.5504
8	0.8611	0.7500	0.7118	0.6058	0.5839	0.5292	0.5363	0.5055	0.5147	0.4905
9	0.8889	0.8026	0.6892	0.5710	0.5772	0.5561	0.5081	0.5147	0.4846	0.5009
10	0.8636	0.7784	0.6764	0.6637	0.5678	0.5341	0.5504	0.4905	0.5009	0.4663

8.6 Selection Bias When the Investigator Assumes PBR in a Study with the BTR

To deal with the wide allocation space of the unequal allocation PBR,
Kuznetsova and Tymofyeyev proposed the BTR [7,8,10], an ARP procedure.
For two-arm allocation, the BTR allows only the allocation sequences that
stay within the unitary squares pierced by the allocation ray on the two-
dimensional grid. This requirement together with the ARP property uniquely
defines the two-arm BTR [8].

Figure 8.4 presents the allocation space for the 7:9 BTR depicted on the
unitary grid. Only the first block is pictured; the allocation space for the

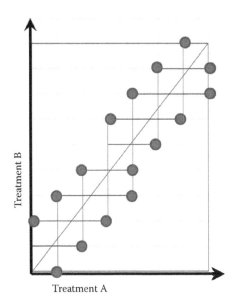

FIGURE 8.4
The allocation space for the 7:9 BTR. Dots: the nodes with the deterministic allocation.

subsequent blocks repeats the one for the first blocks shifted on the grid. The allocation sequences can only follow the sides of the unitary squares pierced by the allocation ray.

The BTR has two nodes in each generation (except the generations at the ends of the blocks: $i = mS$, $m \geq 0$). Similar to a 1:$(S - 1)$ PBR, the BTR allocation is deterministic from one of the two nodes in a generation. For 1:$(S - 1)$ allocation, the BTR coincides with the PBR.

The two-arm BTR is the procedure with the narrowest allocation space and the smallest expected distance from the allocation ray among all ARP procedures with the same allocation ratio [10]. In the open-label single-center trial, this makes the BTR susceptible to the selection bias.

However, the closeness of the BTR sequences to the allocation ray provides an advantage when the investigator uses the directional guessing strategy with the payoff threshold. With the BTR being largely unknown, it is safe to assume that an investigator that faces the allocation ratio of 7:9 will expect the PBR with a block size of 16 to be used in the study. Thus, if the investigator uses the directional strategy with a threshold, even a low threshold would exclude many of the BTR nodes from biasing since they are very close to the allocation ray. In the example of the 7:9 BTR, the threshold $T = 0.2$ excludes all nodes except the four nodes close to the end of the block: (5,8), (7,7), (6,9), and (7,8).

Table 8.8 presents the AESB under the directional strategy with the biasing threshold of 0.15 when the investigator assumes the PBR in a study where the

TABLE 8.8

Average Excess Selection Bias under the Directional Strategy with the Biasing Threshold of 0.15 for $C_1:C_2$ BTR when the Investigator Assumes the PBR

C_1/C_2	2	3	4	5	6	7	8	9	10
1	0.2500	0.2292	0.2375	0.2417	0.2321	0.2366	0.2292	0.2333	0.2364
2		0.1750		0.1393		0.1567		0.1448	
3			0.1339	0.0917		0.1143	0.1059		0.0878
4				0.1069		0.0674		0.0807	
5					0.0750	0.0762		0.0802	
6						0.0627			
7							0.0539	0.0595	0.0532
8								0.0472	
9									0.0420

BTR is used to randomize the subjects. The AESB is provided for a range of allocation ratios $C_1:C_2$, where $0 < C_1 < C_2 \leq 10$ and C_1 and C_2 have no common divisors other than 1. With these allocation ratios, the investigator is likely to assume that the smallest possible PBS is used, and thus we are providing the AESB only for the smallest PBS for each allocation ratio.

Table 8.9 presents the ratio of the AESB when the BTR is used over the AESB when the PBR is used for the directional strategy with the threshold of 0.15. Even with such a small threshold, the use of the BTR provides a notable reduction in the AESB, especially for more balanced allocation ratios. The reduction is more pronounced when the threshold of 0.25 is used (Table 8.10).

Tables 8.8 and 8.9 show that even for a moderate threshold, using BTR instead of the PBR in a study with unequal allocation helps reduce the selection bias when the directional strategy with the threshold is used, especially

TABLE 8.9

Ratio of the AESB under the Directional Strategy with Biasing Threshold 0.15 for $C_1:C_2$ BTR when the Investigator Assumes the PBR over the PBR AESB under the Directional Strategy with the Same Threshold

C_1/C_2	2	3	4	5	6	7	8	9	10
1	1.0000	1.0000	1.0000	1.0000	1.0000	1.0000	1.0000	1.0000	1.0000
2		1.0000		0.8540		0.9240		0.8934	
3			0.8893	0.6906		0.8033	0.7675		0.6945
4				0.8105		0.5976		0.6811	
5					0.6603	0.6765	0.5437	0.7070	
6						0.6073		0.5778	
7							0.5679	0.6305	0.5858
8								0.5374	
9									0.5131

TABLE 8.10

Ratio of the AESB under the Directional Strategy with Biasing Threshold 0.25 for $C_1:C_2$ BTR when the Investigator Assumes the PBR over the PBR AESB under the Directional Strategy with the Same Threshold

C_1/C_2	1	2	3	4	5	6	7	8	9	10
1	1.0000	1.0000	1.0000	1.0000	1.0000	1.0000	1.0000	1.0000	1.0000	1.0000
2		0.8718			0.8540		0.8155		0.8172	
3			0.6818	0.7229			0.6393	0.7165		0.6593
4				0.6099			0.6159		0.5812	
5						0.5362	0.6511	0.6208	0.5650	
6							0.5038		0.6172	
7								0.4900	0.5155	0.5882
8									0.4716	
9										0.4522

when the allocation is not too unbalanced. This helps even more if the investigator biases the allocation only when the next allocation is deterministic.

Of course, in this case, the fact that the BTR is used in the study should not be disclosed to the investigator. This is similar to the practice of not disclosing the PBS in a study protocol or not revealing the block sizes in the random block design [36].

8.7 Discussion

In this chapter, we expanded to unequal allocation the derivations of the selection bias under the Blackwell–Hodges model developed for equal allocation [1,3]. While for equal allocation the selection bias depends on the expected proportion of correct guesses in the whole study, for unequal allocation, it depends on the sum of expected proportions of correct guesses derived separately for treatment groups A and B. Alternatively, the selection bias can be represented through the sum of the within-group differences in the proportions of correct and incorrect guesses.

For a study with unequal allocation, we arrived at a strategy that maximizes the contribution to selection bias due to the next allocation—the exact conditional strategy that does not assume that at the end of the study the treatment group totals are exactly proportional to the allocation ratio. For a large study, where such an assumption provides a good approximation to reality for most restricted allocation procedures, the exact conditional strategy reverts to the directional strategy where the investigator guesses the treatment for which the conditional allocation ratio exceeds the unconditional allocation ratio [2,4].

On the example of the PBR—the most commonly used unequal allocation procedure—we showed that the expected increment to the selection bias factor is low when the achieved treatment group totals indicate the node near the allocation ray and high when the node is far away from it. This provides a rationale to biasing the next allocation only when the payoff to the biasing effort is reasonably high—the directional guessing strategy with a biasing threshold [2]. We tabulated the AESB for the permuted blocks with the dimensions $Q_1:Q_2$, where Q_1 and Q_2 range from 1 to 10, for the directional strategy with and without a threshold to illustrate the impact of the threshold on bias reduction.

And last, we explored the selection bias under the directional strategy with a biasing threshold when the investigator assumes the PBR is used in the study, while in fact the BTR is used. The close proximity of the BTR to the allocation ray means that even for a modest threshold many of the BTR nodes will be considered unworthy of the biasing effort. Thus, the selection bias with the BTR is reduced compared to the selection bias with the PBR. This leads to a recommendation to use the BTR instead of the PBR when there are reasons to believe that the investigator is using the directional strategy with a threshold. The fact that the BTR is used should remain undisclosed to the investigators.

References

1. Rosenberger WF, Lachin JM. *Randomization in Clinical Trials*. John Wiley & Sons, New York, 2002.
2. Berger VW. *Selection Bias and Covariate Imbalances in Randomized Clinical Trials*. John Wiley & Sons, Chichester, 2005.
3. Blackwell D, Hodges JL. Design for the control of selection bias. *The Annals of Mathematical Statistics* 1957; **28**(2): 449–460. doi:10.1214/aoms/1177706973.
4. Berger VW. Quantifying the magnitude of baseline covariate imbalances resulting from selection bias in randomized clinical trials. *Biometrical Journal* 2005; **47**(2): 119–127. doi:10.1002/bimj.200410106.
5. Soares JF, Wu CFJ. Some restricted randomization rules in sequential designs. *Communications in Statistics—Theory and Methods* 1982; **12**: 2017–2034.
6. Zelen M. The randomization and stratification of patients to clinical trials. *Journal of Chronic Disease* 1974; **27**: 365–375.
7. Kuznetsova OM, Tymofyeyev Y. Brick tunnel randomization—A way to accommodate a problematic allocation ratio in adaptive design dose finding studies. *ASA Proceedings of the Joint Statistical Meetings* 2009; 1356–1367.
8. Kuznetsova O, Tymofyeyev Y. Brick Tunnel Randomization for unequal allocation to two or more treatment groups. *Statistics in Medicine* 2011; **30**: 812–824. doi:10.1002/sim.4167.

9. Kuznetsova OM, Tymofyeyev Y. Brick tunnel and wide brick tunnel randomization for studies with unequal allocation. In: Sverdlov A. (ed.). *Modern Adaptive Randomized Clinical Trials: Statistical, Operational, and Regulatory Aspects*: pp. 83–114. CRC Press, Taylor & Francis Group, 2015.

10. Kuznetsova OM. Brick tunnel randomization and the momentum of the probability mass. *Statistics in Medicine* 2015; **34**(30): 4031–4056. doi:10.1002/sim.6601.

11. Kuznetsova OM, Tymofyeyev Y. Expanding brick tunnel randomization to allow for larger imbalance in treatment totals in studies with unequal allocation. *Proceedings of the Joint Statistical Association 2013 Meetings*, Montreal, Canada, 8/4/13 to 8/8/13.

12. Kuznetsova OM, Tymofyeyev Y. Wide Brick Tunnel Randomization—An unequal allocation procedure that limits the imbalance in treatment totals. *Statistics in Medicine* 2014; **33**: 1514–1530. doi:10.1002/sim.6051.

13. Kuznetsova OM. Randomization challenges in adaptive design studies. In: He W, Pinheiro J, and Kuznetsova OM (eds.). *Practical Considerations for Adaptive Trial Design and Implementation*: pp. 157–181. Springer, New York, 2014.

14. Kuznetsova OM, Tymofyeyev Y. Preserving the allocation ratio at every allocation with biased coin randomization and minimization in studies with unequal allocation. *Statistics in Medicine* 2012; **31**: 701–723. doi:10.1002/sim.4447.

15. Kuznetsova OM, Tymofyeyev Y. Expansion of the modified Zelen's approach randomization and dynamic randomization with partial block supplies at the centers to unequal allocation. *Contemporary Clinical Trials* 2011; **32**: 962–972. doi:10.1016/j.cct.2011.08.006.

16. Kuznetsova OM, Tymofyeyev Y. Covariate-adaptive procedures with unequal allocation. In: Sverdlov A. (ed.). *Modern Adaptive Randomized Clinical Trials: Statistical, Operational, and Regulatory Aspects*: pp. 171–198. CRC Press, Taylor & Francis Group, 2015.

17. Proschan M, Brittain E, and Kammerman L. Minimize the use of minimization with unequal allocation. *Biometrics* 2011; **67**(3): 1135–1141. doi:10.1111/j.1541 -0420.2010.01545.x.

18. Kaiser LD. Dynamic randomization and a randomization model for clinical trials data. *Statistics in Medicine* 2012; **31**: 3858–3873. doi:10.1002/sim.5448.

19. Salama I, Ivanova A, Qaqish B. Efficient generation of constrained block allocation sequences. *Statistics in Medicine* 2008; **27**: 1421–1428. doi:10.1002/sim3014.

20. Han B, Enas NH, McEntegart D. Randomization by minimization for unbalanced treatment allocation. *Statistics in Medicine* 2009; **28**: 3329–3346. doi:10.1002 /sim.3710.

21. Hu F, Zhang LX. Asymptotic properties of doubly adaptive biased coin designs for multitreatment clinical trials. *The Annals of Statistics* 2004; **32**(1): 268–301.

22. Sverdlov O, Zhang Y. Statistical properties of covariate-adaptive randomization procedures for trials with unequal treatment allocation ratios. *ASA Proceedings of the Joint Statistical Meetings*, American Statistical Association, Boston, 2014.

23. Titterington DM. On constrained balance randomization for clinical trials. *Biometrics* 1983; **39**(4): 1083–1086.

24. Frane JW. A method of biased coin randomisation, its implementation and its validation. *Drug Information Journal* 1998; **32**: 423–432.

25. Russell D, Hoare ZSJ, Whitaker Rh, Whitaker CJ, Russell IT. Generalized method for adaptive randomization in clinical trials. *Statistics in Medicine* 2011; **30**: 922–934.

26. Lebowitsch J, Ge Y, Young B, Hu F. Generalized multidimensional allocation method. *Statistics in Medicine* 2012; **31**: 3537–3544.

27. Zhao W, Weng Y. Block urn design—A new randomization algorithm for sequential trials with two or more treatments and balanced or unbalanced allocation. *Contemporary Clinical Trials* 2011; **32**(6): 953–961.

28. Ivanova A. A play-the-winner-type urn design with reduced variability. *Metrika* 2003; **58**: 1–13.

29. Berger VW, Ivanova A, Knoll M. Minimizing predictability while retaining balance through the use of less restrictive randomization procedures. *Statistics in Medicine* 2003; **22**: 3017–3028. doi:10.1002/sim.1538.

30. Efron B. Forcing a sequential experiment to be balanced. *Biometrika* 1971; **58**: 403–417.

31. Baldi Antognini A, Giovagnoli A. A new 'biased coin' design for the sequential allocation to two treatments. *Journal of the Royal Statistical Society* 2004; **53**: 651–664.

32. Chen YP. Biased coin design with imbalance intolerance. *Communications and Statistics—Stochastic Models* 1999; **15**: 953–975.

33. Wei LJ. A class of designs for sequential clinical trials. *Journal of the American Statistical Association* 1977; **72**: 382–386.

34. Wei LJ. An application of an urn model to the design of sequential controlled clinical trials. *Journal of the American Statistical Association* 1978; **73**: 559–563.

35. Chen YP. Which design is better? Ehrenfest urn versus biased coin. *Advances in Applied Probability* 2000; **32**: 738–749.

36. Dupin-Spriet T, Fermanian J, Spriet A. Quantification methods were developed for selection bias by predictability of allocations with unequal block randomization. *Journal of Clinical Epidemiology* 2005; **58**: 1269–1276.

9

Unrecognized Dual Threats to Internal Validity Relating to Randomization

Vance W. Berger, Adriana C. Burgos, and Omolola A. Odejimi

CONTENTS

9.1 Introduction 123
9.2 Flawed Trial Evaluations Translate into Flawed Trials 124
9.3 The Propagation of False Knowledge 126
9.4 Chronological Bias versus Selection Bias 128
9.5 Dual Threats to Allocation Concealment 129
9.6 Betting Odds versus Biasing Allocation Only with Certainty 130
9.7 Malice versus Ignorance 131
9.8 Unmasking: Early versus Late 133
9.9 Summary and Conclusions 135
References 137

9.1 Introduction

Medical research is, or should be, conducted with the goal of supporting future medical decisions and policies, so that these can be informed by evidence [1]. Clearly, then, it behooves us, as a society, to ensure that only the best evidence be used to influence future medical decisions. This point may appear obvious on the surface, but the fact that it is not put into practice in the real world means that it may bear repeating, with special emphasis. It is hard to imagine anyone explicitly disagreeing with the statement that good evidence should inform important decisions. But when we insert that extra word "only," we intend to rule out the possibility of flawed evidence also inform important decisions. Safe water should be made available to athletes. ONLY safe water should be made available. In the gray area, if we are uncertain, then we discard the water. We do not say that water should be discarded only if known (with certainty) to be contaminated.

True, we may also say that it is best to hold on to dirty water until we get clean water, and true, medical research tends to cost quite a bit more in time, money, and other resources than water does. But the point remains that we do not want to rely on evidence that may be of questionable integrity. And, unlike water, which we need, we do not technically need to render a definitive judgment on whether or not a given medical intervention is safe and effective. In the absence of valid evidence, we have the option to discard all flawed evidence and declare insufficient evidence to decide one way or the other. However, even *without* valid evidence, we are still hard-pressed to see the merit in relying on flawed evidence. But where do we draw the line? At what point does evidence become sufficiently valid to become useful? This is an important question, and though we cannot answer it definitively in this chapter, we can at least point out some of the flaws to consider when weighing evidence from randomized trials.

In the remainder of this chapter, we shall note briefly how the laissez-faire evaluation of trial quality translates into fatally flawed trials becoming the norm, rather than the exception. Therefore, while there may in fact be some valid trials, the sad reality is that the presumption should be that a trial (conducted according to today's standards) is not valid unless proven otherwise. We shall address how false knowledge, in general, allows this disconnect, fatally flawed trials masquerading as and generally understood to be rigorous and beyond reproach, to continue. After that, we shall focus on several specific misunderstandings that relate to randomization methods, allocation concealment, and masking, and try to clear them up.

9.2 Flawed Trial Evaluations Translate into Flawed Trials

It is generally recognized that randomized clinical trials represent the best form of medical evidence, either alone or in combination for systematic reviews. Therefore, when trials are available, it is hard to imagine what benefit accrues from mixing in observational studies that are subject to more biases [2]. But are all trials of uniform methodological quality? In fact, they are not. It is rather unfortunate that the scrutiny of these randomized trials leaves quite a bit to be desired, and tends to use broad brushes to produce the mistaken impression that in fact all trials are of not only equal, but also uniformly superb, methodological quality.

Perhaps the worst instance of this trend is the use of the grossly incomplete Jadad score to evaluate trial quality, with its reduction of a complex trial into a caricature, asking only a ridiculously small number of questions, with generous (and highly inappropriate) partial credit awarded just for playing. For example, the mere claim of randomization earns a point, even without any substantiation. The Jadad score has been amply discredited in the literature [3–5], and would be patently absurd even without its flaws being pointed out (since they are so obvious), yet, indefensibly, the Jadad score *still* remains in wide use in practice because it is tantamount to not evaluating trial quality at all. This most forgiving "evaluation" (if we can call it that), which might most accurately be described as an "anything goes" participation award, is exactly what one would want when either (1) covering up known trial flaws or (2) hoping to do as little work as possible. After all, even a researcher *not* trying to bias a trial may still balk at having to use research methods that differ substantially from those used in prior trials (inertia).

The result is that fatally flawed trials can fly under the radar and masquerade as rigorous trials, and the public is led to believe that faith in the system of medical research is warranted. In fact, it isn't. As a result of sham trial evaluation, with a wink and a nod, (1) trials, taken as a whole, are conducted to appallingly low standards and (2) the ubiquitous flaws in randomized trials are rarely recognized as such. We might add in (3) even in those rare cases when the flaws *are* recognized, this information tends to be noted and shelved, rather than acted upon in any meaningful way (reminiscent of that recent TV commercial in which a security *monitor* merely monitors for crimes, but then does nothing but take note when a crime actually does occur). Though the rationale for this policy is rarely articulated explicitly, it would seem to be rooted in the misguided belief that randomized trials are, merely by virtue of *being* randomized trials, beyond reproach.

The reality of the situation is, as previously noted, that trials are consistently conducted to appalling low methodological standards, while simultaneously being conducted to exacting standards of procedure and precedent that lend the appearance of validity but in fact have nothing to do with validity or methodological quality. As long as all the right forms are signed in triplicate, in black ink, and as long as industry standards are followed, it matters little if the trial designs (or the standards themselves) are actually any good. In other words, the sham applies not only to the evaluation of trial quality but also to the trials themselves. It is for this reason that Winston Churchill's famous quote of November 11, 1947, was modified so as to recognize randomized trials as being the worst possible medical study design, except for all the others [6].

And this is not a tempest in a teapot. The flaws that we see in trials on a regular basis, even those heralded as being rigorous and published in the most prestigious journals, are generally sufficient to invalidate the trials; there is no compensation for getting other aspects right [7]. Precedent has already been mentioned as one culprit contributing to this problem, and indeed inertia is a major contributor to the scourge that is flawed medical research [8]. But, alas, it is not the *only* problem. Just as distorted evidence from fatally flawed trials flows like a polluted river into the ocean that is medical knowledge, so too does distorted understanding propagate itself, one generation to the next, in misguided instruction, both in university settings and in less formal settings. The wrong messages are taught in the schools. But this is not haphazard.

9.3 The Propagation of False Knowledge

The exact *same* wrong messages are reinforced over and over again, in seminars and conferences, on journal pages, in textbooks, and informally in *ad hoc* discussions. As a result, researchers tend to "know" facts that are simply not true, and they can take solace in knowing that others also "know" these same facts. Belonging to the club or jumping on a bandwagon can be a rather powerful incentive, even among those who should know better and who should be above giving in to such pressures.

In 1853, Ernestine L. Rose noted that "a child may be made to believe a falsehood and die in support of it, and therefore there can be no merit in a [mere] belief." A similar sentiment was stated by Mark Twain; "It ain't what you don't know that gets you into trouble, it's what you know for sure that just ain't so." As an analogy, I recall, as a kid watching cartoons, seeing Daffy Duck get shot, and the end result was merely an annoyance. He had only to dust himself off and straighten himself out and he was as good as new. I saw this with my own two eyes, and it never occurred to me at the time to question the veracity of what I was being led to believe. Therefore, I "knew" as a fact that getting shot was annoying, but not much more than that, and at times I felt the need to correct the silly grown-ups who had no clue and actually thought of gun violence as a major problem. This anecdote will hopefully clearly illustrate the fact that strength of belief should never be accepted as a metric of validity, not when a kid pontificates on the results of being shot, and not when a misguided researcher pontificates on how to conduct a trial, either in general, or on how to randomize these trials, in particular.

How many researchers "know" for sure that unrestricted randomization is a solution to the problem, as has been recently proposed [9]? How many researchers "know" that permuted block randomization is a viable solution? How many "know" that we are OK if only we vary the block sizes? How many "know" that we are OK as long as the trial is masked, or as long as we operate with allocation concealment? Each of these classes of misguided and falsely enlightened individuals will go on to conduct trials in accordance with their own distorted understanding, and then the trials themselves go on to produce distorted results, bearing little resemblance to the reality governing the situation. This represents tangible harm to the patients who depend on medical researchers to produce results that can be expected to bear some resemblance to reality.

It is also true that falsehoods are most convincing when they contain at least an element of truth in them. Therefore, the most harm comes from the falsehoods that are plausible, because these are believed, and then taken as gospel, whereas the sillier ones tend to be dismissed out of hand before they can do any harm. Some of the falsehoods that are widely accepted as true (some already noted above) would include the following:

1. Randomization precludes the possibility of selection bias.
2. Randomization coupled with masking precludes the possibility of selection bias.
3. Randomization coupled with allocation concealment and masking precludes the possibility of selection bias.
4. Randomization ensures that any baseline imbalances are necessarily random.
5. Unrestricted randomization precludes the possibility of selection bias.
6. The best way to randomize is with permuted blocks.
7. The best way to randomize is with unrestricted randomization.
8. Varying the block sizes renders advance allocation prediction impossible.

Each one of these statements can be supported with a compelling argument. And each one is demonstrably false. This bold statement may require some clarification. Yes, it is true that randomization coupled with *perfectly successful* masking will preclude the possibility of selection bias, but when can masking ever be demonstrated to be perfectly successful? What kind of demonstration would suffice to establish that allocation concealment was successful? The claim always means that the effort was made, and not that the effort was successful. Therefore, neither #2 nor #3 is true in any meaningful sense. Likewise for #5, since, as argued in another chapter, unrestricted randomization is never actually used anyway, even when it is claimed, and what *is* used when unrestricted randomization is claimed certainly cannot eliminate selection bias.

Moreover, each one of these statements also represents tangible harm to real patients, because the misguided belief in these statements leads to research practices that practically ensure bias in the trials that are conducted on behalf of medical patients. Each one of these statements has been debunked in the relevant literature. And, for the most part, that literature has been ignored, and fatally flawed randomization methods continue to be used as if some or all of these statements were true. The distorted understanding of the researcher is transferred to the trial itself, which then becomes contaminated, so to speak, and may no longer be valid in any meaningful sense.

This failure to use best practices represents a dereliction of duty that borders on research malpractice, and this remains true even though, as noted, these decisions can be defended by misguided arguments that at least *appear* to be correct. The problem is that we often face a dilemma, in the form of two simultaneous threats, either one of which can be adequately addressed in isolation. But addressing both threats simultaneously is much more difficult, and, as a result, many researchers simply cannot be bothered to even try. Instead, they take the easy way out, by addressing only one horn of the dilemma (one of the two threats), without even acknowledging the other one. In the remainder of this chapter, we shall illustrate some of these dilemmas and dual threats, and discuss the misguided manners in which they are usually handled, and offer improvements for handling them better in future clinical trials.

9.4 Chronological Bias versus Selection Bias

The failure, in general, of trials to use best practices regarding randomization often comes down to a failure to recognize dual threats, and to instead focus on a single threat to the exclusion of the other threat. In fact, there are so many dual threats, sometimes one embedded within another, that the overall structure may even resemble a fractal. As one example, we want randomization to successfully address the dual threats of (1) chronological bias [10] and (2) selection bias [11,12]. Consideration of only the first threat, chronological bias, seems to be the most common approach in practice, although it is noteworthy that this seems to be done most often at a subconscious level. The most common remedy for chronological bias is permuted block randomization, and this is what researchers use almost universally, but without ever stating, or likely without even knowing, why they are doing so. Nor would they recognize that the solutions to chronological bias, permuted block randomization and alternation, do not address selection bias at all. Consideration of only the second threat, which is less common, likewise leads to a solution that appears to be adequate, namely, the same unrestricted randomization that we have criticized in Chapter 4 of this book, and also in Ref. [13].

Permuted block randomization has been amply discredited in the literature, time and time again [14–20], and yet this *still* remains the industry standard. In fact, so ubiquitous is permuted block randomization that it often just goes without saying, and so journals often do not even ask for (or demand) clarification when authors provide no specifics whatsoever regarding how they randomized. Perversely, no explanation is needed when one wants to use the worst possible randomization method, and yet any attempt to improve on this and use a randomization method that is actually valid will expose one to interrogation.

The situation here bears some resemblance to the one governing the standard QWERTY keyboard, and the alleged (although at times disputed) superiority of the Dvorak keyboard. For our purposes, it is of no consequence that some have suggested that the QWERTY is actually superior. The fact remains that it is generally held (at least by those familiar with both) to be far inferior to the newer Dvorak keyboard, which is never used only because it was second on the scene. Whether or not path dependence is at work in keeping the QWERTY in play, it is *certainly* at work in keeping permuted block randomization in play, where there is no analogous suggestion of superiority to or even equivalence to the vastly superior MTI randomization procedures, which address both horns of the dilemma [11,12,14], chronological bias and selection bias, but just happened to have the bad luck of coming along later, after permuted block randomization was already entrenched as the standard.

Alas, the stakes are much higher in clinical research than they are in the typing world, and the patients who depend on medical researchers to use best practices (as opposed to relying exclusively on social proof, as so many seem to do now) get cheated, often with grave consequences, when trials are randomized with permuted blocks.

This may sound like an outlandish claim. How can something so inconsequential as the randomization procedure actually harm patients? But in fact this is not far-fetched at all. Flawed randomization leads to unbalanced treatment groups, which in turn lead to distorted trial results. It is then these distorted results, and not the results that would best reflect the reality of the situation, that go on to inform future medical practice. If a safe and effective treatment is found to be useless, then patients are harmed by not knowing that this treatment can help them. If, on the other hand, an ineffective treatment is found to be safe and effective, then again real patients are harmed, as they will be led to this treatment that cannot help them, and may well harm them. Either way, distortions in the research record produce tangible harm, and these distortions are in turn caused by flawed research methods of all varieties, including the aforementioned nearly ubiquitous permuted block randomization procedure.

In short, using any method that does not represent best practices cannot be defended, and so it is high time that we, as a field, move on from permuted block randomization to the far superior MTI randomization methods. Kaizen, and not inertia, is what is needed. Moreover, it does seem quite incongruous to pass off medical research to an unsuspecting public as being pristine, and conducted by experts who know what they are doing, when in fact, at least with regard to randomization, more often than not they simply use the only method they have heard of. This is conduct unbecoming of a medical researcher, and should no longer be tolerated.

9.5 Dual Threats to Allocation Concealment

The threat of selection bias is of course addressed by allocation concealment, because the mechanism of action, so to speak, involves the advance knowledge of future allocations. But allocation concealment itself faces dual threats, namely, (1) the direct observation of (all or part of) the allocation sequence and (2) its prediction by exploiting known patterns coupled with knowledge of at least some of the prior allocations [21,22]. Once again, the general trend is to fixate perversely on the first threat, while conveniently ignoring the second one. This leads to the absurd notion that successful allocation concealment results whenever one simply makes the claim, as if it were just a box to check off on a list. We hid (or, more correctly, *attempted* to hide) the allocation sequence, so therefore no encryption is needed.

As an analogy, we recall that, in September of 1862, Robert E. Lee issued Special Order #191 to the Army of Northern Virginia, outlining his plans to split his forces to try to encircle George McClellan's Union Army of the Potomac. This order was obviously not meant to be seen by the Union soldiers, and then, as now, the mere effort alone was confused with the success of that effort. Since we do not intend to show this order to the Union Army, we have no reason to encrypt it. Alas, the order did in fact end up in Union hands (along with a few cigars), and now the Union Army knew exactly what the Confederate Army was planning, and had ample time to formulate a strategy to counter this plan. Of course, the Army learned an important lesson here, and by World War II, 80 years later, encryption was already an integral part of the overall military strategy [23]. When will we learn this same lesson in planning how to randomize?

In point of fact, allocation concealment is *not* ensured merely by the effort (and even less so by the claim without any effort at all). Any unsubstantiated claims of allocation concealment must therefore be viewed with suspicion commensurate with the suspicion one would show a soldier who guaranteed that the enemy would never see these plans and then supported this statement by saying only that we kept these plans out of enemy hands (meaning, *intended to*). The reality is that completely successful allocation concealment is precluded by the combination of *any* restrictions on the randomization and any unmasking (whether intentional or not, whether known or not). Yet we still need some restrictions, for the reasons already mentioned in Chapter 4, dealing with the problems inherent in the use of unrestricted randomization. The key is to maximize allocation concealment (recognizing that it is not a binary phenomenon) by selecting the right restrictions prudently [11,12]. Herein lies the basis for preferring MTI randomization to permuted blocks or unrestricted randomization [14].

9.6 Betting Odds versus Biasing Allocation Only with Certainty

When there is the combination of unmasking of any of the allocations coupled with any restrictions on the randomization, we again face dual threats, namely, investigators who will select patients strategically (1) only when they are certain of the next allocation or (2) whenever there are any betting odds, which is to say, any conditional allocation probabilities that differ, by any amount, from their unconditional counterparts. Consideration of only the first threat leads to the misguided suggestion that we continue using permuted blocks, but simply vary the block size. Consideration of only the second threat leads to the promotion of the big stick [24], which actually is not a bad randomization procedure at all. However, its optimality is also easily overstated.

In fact, varying the block sizes will not even necessarily eliminate all deterministic allocations [12,25], despite the fact that so many "know" this to be the case. Consider, for example, varying the blocks among sizes two and four, in an entirely unmasked trial. Since the largest block size is four, the induced MTI is two. That is to say that the treatment group sizes can never differ by more than two. If anywhere during the trial we notice four consecutive allocations to the same treatment group, then we know with certainty that balance was restored after the first two of these, and the last two took the allocation sequence to the MTI boundary. The next two allocations will both be deterministic at this point, as we are necessarily at the midpoint of a block of size four.

But this is not even the primary objection to using varied block sizes. The far more serious consideration is that this approach allows for quite a bit of prediction that falls short of deterministic allocations, much more, in fact, than would be the case with a fixed block size equal to the largest block size considered for the scheme with varied block sizes. In other words, one can predict equally well whether there is a fixed block size of four or varied block sizes of two and four, but these predictions, utilizing the convergent strategy of guessing whichever group has been allocated less often so far, will be correct more often if the block sizes are varied [12]. In summary, varying the block sizes is somewhat effective, but not nearly as much as generally thought, at eliminating deterministic allocations, yet still remains an inappropriate randomization procedure in practice because of its dismal failure at curtailing predictable allocations. Conversely, the big stick procedure will tend to minimize the number of predictable allocations, but any allocation that *is* predictable when using the big stick will also be deterministic. It does not fail often, but when it does, it fails spectacularly. This is not to say that the big stick should not be used. It is a good procedure, with much merit to recommend it. Still, the maximal procedure seems more appropriate in how it balances these two threats [26].

9.7 Malice versus Ignorance

It is worth considering the dual threats posed by (1) those intending to distort the trial results for personal gain and (2) those who do so inadvertently, because the remedial measures one would take to prevent and correct for such behavior will differ across the two. It is popular to take the position that nobody would ever intentionally engage in an unethical practice, but time and time again we see that this is simply not true, not only in general terms but also specifically in the context of medical research. See, for example, Ref. [8], Chapter 3 of Ref. [12], and Refs. [27–30].

Moreover, even if the literature were not already replete with well-documented examples of investigators trying to game the system for personal benefit, we would *still* be on rather shaky ground in using faulty induction to declare that since (1) no examples of malfeasance are known, (2) there never has been malfeasance and (3) therefore there never will be any malfeasance, and all investigators, past, present, and future, are all entirely ethical. Clearly, neither leap of faith would be justified. First, being unaware of a specific phenomenon does not render this phenomenon nonexistent, especially when there is a strong incentive to remain willfully ignorant of its existence. And second, even if it could somehow be demonstrated that to this point no trials have even been manipulated, that still would not be sufficient to declare with any credibility that all future trials will follow suit. The recognition that there is a first time for everything is, after all, the foundation for the entire insurance industry. Imagine if everyone considered auto or health insurance to be unnecessary on the basis of never before having been in a car accident or hospitalized, respectively.

In fact, Pearl Harbor might have been spared with more enlightenment along these lines. As it happens, nine years earlier, on February 7, 1932, Pearl Harbor was decimated in a simulated attack that ultimately was emulated on December 7, 1941. The vulnerability was clearly demonstrated by Harry E. Yarnell, in hopes that defenses would be built to shore up this vulnerability. Instead, because such an attack had never happened before, the lesson was ignored by those who should have known better, and studied thoroughly by those who *did* know better, and used it to their advantage when attacking. Pearl Harbor has never been attacked before; therefore, it never will be. No trial has ever been manipulated before (this is where the parallel ends; in point of fact many trials *have* been manipulated, but this fact carries no weight among those who choose to ignore it), so therefore no trial ever can or will be manipulated.

So while a much stronger statement can easily be supported with available evidence, for our purposes, it suffices to state only that it is in the realm of possibility that some future trials could be manipulated intentionally, and also that some others (or even the same ones) could also be fatally flawed through ignorance alone. Neither possibility can be ruled out with any credibility. Consideration of only the first of these threats leads to inadequate solutions, and lends itself to thinking along the lines of "Well, the only way for a trial to be manipulated is with malfeasance, and that is a rather strong accusation that I am not prepared to level at this time, and since I am not aware of any problem, therefore there is no problem" (especially when this statement is made by an individual with a cozy relationship with the investigators he or she should be monitoring). We can remedy this problem with mandatory disclosure of conflicts of interest in research reports. This is not to say that those dishonest investigators will jump at the opportunity to disclose incriminating information but rather that if the expectation is to have conflicts of interest disclosed, any reported findings and discussion that lack this requirement will be seen as incomplete and unprofessional.

This response to the suggestion that diligence be applied in scrutinizing the trial for biases fails miserably in two distinct ways. First, as noted, absence of evidence is not always evidence of absence. If there is a bias caused by bad intentions, then we can hardly expect the perpetrators to advertise this fact. And second, even if we did know that there was no willful bias in this trial, it still would not mean that there was no inadvertent bias (caused by using improper research methods without intending to do so).

Consideration of only the second threat again leads to two failures, first in that it summarily discounts the possibility of malice and second in that it tends to minimize the consequences of honest mistakes. It follows the misguided notion that we are all adults here, all trying to do the right thing, and sure, some mistakes may be made, but these are honest mistakes, and therefore are harmless. Nothing could be further from the truth. In no way do we wish to trivialize the harm done by those intent on altering trial results to suit their own purposes, but an argument can be made that this is the lesser of two evils, especially if we believe that, while some do engage in this unethical behavior, they at least represent a relatively small proportion of all trial researchers. So along these lines, we echo Dr. Martin Luther King in noting "Shallow understanding from people of good will is more frustrating than absolute misunderstanding from people of ill will."

Moreover, reaching out to those who engage in selection bias without bad intentions can help create a research atmosphere that is more hostile to those who would do so intentionally. If we entertain the idea that this problem may just be under-education of testing methods by both malicious and ignorant investigators, we can clearly see that refined education shall "set us free" (so to speak). When we include greater trial scrutiny and sanctions for research malpractice, we avoid dead-end solutions and instead find accountability for the process of reputable clinical trials.

9.8 Unmasking: Early versus Late

If certain allocations are unmasked inadvertently, then a useful distinction may be drawn between early and late unmasking. Though different definitions are possible, for our purposes, a rather crude classification suffices, and unmasking is considered to be early if it occurs before the next allocation, and late otherwise. Crisp [31] pointed out (correctly) that in the case of late unmasking only, our best bet is to use permuted block randomization with small block sizes. This is because the only opportunity to predict future allocations based on past ones, at least when permuted blocks are used, is *within* the block. No matter how many prior allocations are known, the first allocation of each block will have successful allocation concealment, and cannot be predicted. Moreover, if this first allocation within a block remains successfully masked, then the next allocation in that same block will also enjoy successful allocation concealment, independent, once again, of how many prior allocations from earlier blocks have already been unmasked.

With this in mind, we recognize that we are exposed to the danger of prediction only while a block remains opened. More accurately, we are exposed to the danger that comes from unmasking only while the block in which the unmasked allocation was rendered remains open. Therefore, we want to minimize our exposure by closing blocks as quickly as possible after opening them. This is accomplished, of course, with the smallest blocks possible, which, with two treatment groups, would be blocks of size two. This analysis establishes the benefit of smaller blocks relative to larger ones, at least with regard to late unmasking. But how would blocks of any size compare to MTI procedures? In fact, for the purposes of this analysis, MTI procedures might be considered to be one large block, with no statute of limitations. There is no wall of separation between even the earliest allocations and the latest ones. So by this consideration, MTI procedures would be worse than permuted blocks of any block size.

Therefore, if one were to assume that all unmasking were late, then the ideal remedy would be to use permuted block randomization with the smallest possible block size of two (or, more generally, with the block size matching the number of treatment groups, at least for trials with equal allocation). But how good a solution is this when we broaden the perspective and consider the dual threat, namely, that some unmasking can also be early? We have already discussed the fatal flaws inherent in permuted block randomization, and this still applies as long as any of the unmasking is early. In fact, contrary to what we found when all unmasking was late, with early unmasking we find that the smaller the block size, the more influential one unmasked allocation is for all future allocation in that same block. So, as is often the case, and as is the general theme of this chapter, there is a trade-off, and we must defend against both threats.

This does, of course, raise the question of whether unmasking could actually be early. To address this question, we note that injection site reactions would be one example of early unmasking [32]. Another would be an identifiable difference in taste or smell, as was the case in the famous 1975 NIH trial of ascorbic acid (vitamin C) [33,34]. Therefore, early unmasking cannot be dismissed out of hand, and the solution to late unmasking (randomization with small block sizes) fails because it does not protect against selection bias in the case of early unmasking. True, MTI randomization is worse than permuted blocks with small block sizes *if all unmasking is late*, but here the magnitude of difference is not so great. In an overall sense, then, MTI randomization is still to be preferred to permuted blocks (and by a large margin), and large blocks (or MTI values) are still preferred to smaller ones (again, by a large margin).

9.9 Summary and Conclusions

In this chapter, we have discussed several dual threats to successful allocation concealment, and, therefore, to trial validity. Each of these dual threats is characterized by the asymmetric attention paid in general to one threat at the expense of the other. Table 9.1 summarizes these dual threats, indexed by section number (S in the table). We see that (as already noted) in each case there are two related yet distinct problems, one of which tends to be more recognized in general, and that the solution to this more recognized one tends to fail miserably as a remedy for the unrecognized problem. If a building to be secured could be entered through either a primary front door or a secondary back door, and if we had two armed guards to secure this building, then would we station both of them at the front and leave the rear completely unguarded? That is exactly what tends to be done when ad hoc approaches to these dual problems are used in practice, as these ad hoc approaches tend to address only one of the two threats.

As it turns out, there is some convergence of the ideas, as more than one of the problems considered points to the need for MTI randomization instead of the obsolete and inappropriate permuted blocks randomization method.

TABLE 9.1

Unrecognized Dual Threats and Omnibus Solutions

S	Recognized	Solution 1	Unrecognized	Solution 2	Overall Solution
4	Chronological bias	Permuted blocks	Selection bias	Unrestricted	MTI methods
5	Sequence observed	Allocation concealment	Predicted	Unrestricted	MTI methods
6	Deterministic	Varied block sizes	Betting odds	Big stick	Maximal procedure
7	Malice	Disclosure of conflicts of interest	Ignorance	Refined education	Refined education and sanctions
8	Late unmasking	Small permuted blocks	Early unmasking	Large permuted blocks	Large MTI value

The primary MTI methods are the big stick [24], Chen's procedure with any selected biasing probability [35], the maximal procedure [11,12,14], and the asymptotic maximal procedure [26]. Any one of these can be justified as an appropriate randomization method. But Section 9.6 suggests that the maximal procedure may be the best of all of them (more insight here can be gained from the relevant references), and Section 9.8 shows that we prefer to use a large MTI value in conjunction with an MTI procedure. These are the primary take-home messages, with regard to preventing selection bias in randomized trials.

But selection bias is but a drop in the ocean that is the set of biases that plague medical studies routinely, and so we also found it appropriate to touch upon the overly permissive context in which these biases occur. Some of what we offered here is certainly relevant to improving the methodological quality of medical studies in general, and not limited to the specific issue of selection bias and randomization.

Section 9.7 discussed two classes of researchers that we are hoping might benefit from this presentation. There is the researcher who is uninformed or misguided about trial research methodology and is unaware of the literature now condemning certain techniques that they may have been relying upon for years. Then there are researchers who, for personal gain, will not wish to use more valid methods precisely because they *do* know that the obsolete methods they are currently using will allow them to manipulate the trial results for their own purposes. Either way, flawed methodology renders a trial vulnerable to bias. This would not be a problem if sufficient detection systems were in place to root out and eliminate biases and faulty research methods. Unfortunately, this is not the case. The existence of equally flawed evaluation systems almost ensures that valid evaluations are never conducted, and so the problem continues.

For example, the Jadad scoring system is grossly inadequate for its purpose of evaluating trial quality, and its shortcomings are readily apparent; no special effort or brilliance is needed to recognize these shortcomings. Yet it remains in frequent use, presumably by those who wish to hide behind the plausible deniability of not knowing better, but likely also by those who know full well that using the Jadad score will allow fatally flawed trials to appear to be rigorous. The same might be said for permuted block randomization, whose flaws may be a bit less obvious than those of the Jadad score, but this is countered by the fact that they have also been discussed (and criticized) much more often in the literature. So ignorance can no longer be accepted as a valid excuse here, either. And yet, this flawed method, too, is used all the time.

The fact that this intolerable situation persists raises some rather uncomfortable questions. How can this be allowed to continue? Who would be in the best position to put a stop to it? Who benefits from the status quo? Who is harmed by it? If a researcher actually does put forth the effort to use nothing but best practices, then will this help or hurt his or her career? The answers to these questions are readily apparent, and must form the basis of any rational plan to improve the situation. Clearly, there is more than ignorance at play here, and malice must be dealt with too. But even here, a general ignorance (on the part of those surrounding the party intentionally biasing trials), coupled with a lack of credulity regarding the extent of this malice, also serves to allow the malice to fly under the radar, and continue. Therefore, while addressing ignorance may not suffice all by itself, it certainly is a very good start, and a major step in the right direction. We tried to address multiple manifestations of ignorance by enumerating and discussing several widely held misunderstandings all pertaining to randomization in clinical trials.

When we consider all the aforementioned misunderstandings that permeate even just this one little corner of clinical trial research, we see that what we need most is education. And there is no shortage of medical education, but for all the opportunities that exist for individuals to gain an education in this field, the field as a whole has failed, miserably, to educate the set of researchers (considered as a whole) properly on these crucial elements of trial design. Therefore, the status quo is not good enough, and the curriculum is clearly in dire need of being revamped and brought up to date. This chapter (along with the rest of the book) represents an effort in this direction, but of course much more will be needed in the future. We look forward to these efforts.

References

1. Hodge, J.G., Jr. (2005). An enhanced approach to distinguishing public health practice and human subjects research. *Journal of Law, Medicine & Ethics*, 33(1), 125–141. doi: 10.1111/j.1748-720X.2005.tb00215.x
2. Berger, V.W. (2013). What do non-randomized trials offer above and beyond randomized trials? *Contemporary Clinical Trials*, 35(1), 168–169. doi: 10.1016/j.cct.2013.03.008
3. Berger, V.W. (2006). Is the Jadad score the proper evaluation of trials? *Journal of Rheumatology*, 33(8), 1710.
4. Palys, K. & Berger, V.W. (2013). A note on the Jadad score as an efficient tool for measuring trial quality. *Journal of Gastrointestinal Surgery*, 17(6), 1170–1171. doi: 10.1007/s11605-012-2106-0
5. Alperson, S. & Berger, V.W. (2013). Beyond Jadad: Some essential features in trial quality. *Clinical Investigation*, 3(12), 1119–1126. doi: 10.4155/cli.13.103
6. Berger, V.W. (2011). Trials: The worst possible design (except for all the rest). *The International Journal of Person Centered Medicine*, 1(3), 630–631. doi: 10.5750/ijpcm.v1i3.126
7. Palys, K., Berger, V.W. & Alperson, S. (2013). Trial quality checklists: On the need to multiply (not add) scores. *Clinical Oral Investigations*, 17(7), 1789–1790. doi: 10.1007/s00784-013-1020-5

8. Berger, V.W. (2015). Conflicts of interest, selective inertia, and research malpractice in randomized clinical trials: An unholy trinity. *Science and Engineering Ethics*, 21(4), 857–874. doi: 10.1007/s11948-014-9576-2

9. Kahan, B.C., Rehal, S. & Cro, S. (in press). Risk of selection bias in randomised trials. *Trials*. Retrieved from https://trialsjournal.biomedcentral.com/articles/10.1186/s13063-015-0920-x

10. Matts, J.P. & McHugh, R.B. (1983). Conditional Markov chain designs for accrual clinical trials, *Biometrical Journal*, 25(6), 563–577.

11. Berger, V.W., Ivanova, A., & Deloria-Knoll, M. (2003). Minimizing predictability while retaining balance through the use of less restrictive randomization procedures. *Statistics in Medicine*, 22(19), 3017–3028. doi: 10.1002/sim.1538

12. Berger, V.W. (2005). *Selection Bias and Covariate Imbalances in Randomized Clinical Trials*. Chichester, England: John Wiley & Sons.

13. Berger, V.W. (2016). Risk of selection bias in randomized trials: Further insight. *Trials*, in press. doi: 10.1186/s13063-016-1597-5

14. Berger, V.W., Agnor, R.C. & Bejleri, K. (2016). Comparing MTI randomization procedures to blocked randomization. *Statistics in Medicine*, 35(5), 685–694. doi: 10.1002/sim.6637

15. Zhao, W., Weng, Y., Wu, Q. & Palesch, Y. (2012). Quantitative comparison of randomization designs in sequential clinical trials based on treatment balance and allocation randomness. *Pharmaceutical Statistics*, 11(1), 39–48. doi: 10.1002/pst.493

16. Zhao, W. & Weng, Y. (2011). Block urn design—A new randomization algorithm for sequential trials with two or more treatments and balanced or unbalanced allocation. *Contemporary Clinical Trials*, 32(6), 953–961. doi: 10.1016/j.cct.2011.08.004

17. Zhao, W. (2014). A better alternative to stratified permuted block design for subject randomization in clinical trials. *Statistics in Medicine*, 33(30), 5239–5248. doi: 10.1002/sim.6266

18. Zhao, W. (2013). Selection bias, allocation concealment and randomization design in clinical trials. *Contemporary Clinical Trials*. 36(1), 263–265. doi: 10.1016/j.cct.2013.07.005

19. Zhao, W. (2016). A better alternative to the inferior permuted block design is not necessarily complex. *Statistics in Medicine*, 35(10), 1736–1738. doi: 10.1002/sim.6858

20. Kennes, L.N., Cramer, E., Hilgers, R.D. & Heussen, N. (2011). The impact of selection bias on test decisions in randomized clinical trials. *Medical Journal of Australia*, 30(21), 2573–2581. doi: 10.1002/sim.4279

21. Berger, V.W. (2005). Is allocation concealment a binary phenomenon? *Medical Journal of Australia*, 183(3), 165.

22. Berger, V.W. & Do, A.C. (2010). Allocation concealment continues to be misunderstood. *Journal of Clinical Epidemiology*, 63(4), 468–470. doi: 10.1016/j.jclinepi.2009.09.004

23. Kahn, D. (1991). *Seizing the Enigma: The Race to Break the German U-Boat Codes, 1939–1943*. Boston: Houghton Mifflin.

24. Soares, J.F. & Wu, C.F.J. (1983). Some restricted randomization rules in sequential designs. *Communications in Statistics Theory and Methods*, 12(17), 2017–2034. doi: 10.1080/03610928308828586

25. Berger, V.W. & He, X. (2015). Concealing block sizes is not sufficient. *Clinics in Orthopedic Surgery*, 7(3), 422–423. doi: 10.4055/cios.2015.7.3.422

26. Zhao, W.L., Berger, V.W. & Yu, Z. (2016). The asymptotic maximal procedure for subject randomization in clinical trials. *Statistical Methods in Medical Research.* doi: 10.1177/0962280216677107

27. Moynihan, R. & Cassels, A. (2005). *Selling Sickness: How the World's Biggest Pharmaceutical Companies Are Turning Us All into Patients.* Avalon, New York: Allen & Unwin.

28. Abramson, J. (2004). *Overdosed America: The Broken Promise of American Medicine.* New York: Harper Collins.

29. Angell, M. (2005). *The Truth about the Drug Companies: How They Deceive Us and What to Do about It.* New York: Random House.

30. Welch, H.G., Schwartz, L. & Woloshin, S. (2011). *Overdiagnosed: Making People Sick in the Pursuit of Health.* Boston, Massachusetts: Beacon Press.

31. Crisp, A. (2016). On the limitations of permuted blocked randomization. *Statistics in Medicine*, 35(12), 2109–2110. doi: 10.1002/sim.6756

32. Berger, V.W. & Agnor, R.C. (2016). Delayed unmasking and selection bias: Reply to Crisp. *Statistics in Medicine*, 35(12), 2111–2112. doi: 10.1002/sim.6799

33. Karlowski, T.R., Chalmers, T.C., Frenkel, L.D., Kapikian, A.Z., Lewis, L.L., Lynch, J.M. (1975). Ascorbic acid for the common cold: A prophylactic and therapeutic trial. *JAMA*, 231(10), 1038–1042. doi: 10.1001/jama.1975.03240220018013

34. Hemilä, H. (1996). Vitamin C, the placebo effect, and the common cold: A case study of how preconception influence the analysis of results. *Journal of Clinical Epidemiology*, 49(10), 1079–1087. doi: 10.1016/0895-4356(96)00189-8

35. Chen, Y.P. (1999). Biased coin design with imbalance tolerance. *Communications in Statistics*, 15, 953–975.

10

Testing for Second-Order Selection Bias Effect in Randomised Controlled Trials Using Reverse Propensity Score (RPS)

Steffen Mickenautsch and Bo Fu

CONTENTS

10.1 Introduction.. 141
10.2 Materials and Methods 143
    10.2.1 Trial Simulation and Parameters.. 143
    10.2.2 Trial Scenarios 144
    10.2.3 Bias Testing 145
    10.2.4 Data Analysis and Summary Measures.................................. 146
10.3 Results 146
    10.3.1 Test Sensitivity and Specificity 146
    10.3.2 Trial Parameters and Test Accuracy.. 149
10.4 Discussion.. 149
    10.4.1 Study Method Limitations... 149
    10.4.2 Discussion of Test Results... 150
    10.4.3 Recommendation for Further Research.................................... 154
10.5 Conclusion 155
Conflict of Interest Statement 155
References.. 155
Appendix File 1 Four-Step RCT Simulation with R code "rps.gen" 156
Appendix File 2 Test Results (Raw Data) 157

10.1 Introduction

Selection bias affects the internal validity of clinical trials [1] by favouring one clinical intervention above others [2]. This bias is introduced into non-randomised controlled trials when knowledge of certain participant characteristics, known to be conducive to the success of one particular intervention, and knowledge regarding the allocation of such patients into compared intervention groups are applied together.

The generation of a random allocation sequence in randomised controlled trials (RCTs) may provide an equal chance for trial subjects to be allocated into either intervention group. However, such allocation may be subverted (i) when trial subjects are selected before the generation of the allocation sequence, (ii) when this sequence is directly discovered or (iii) when correctly predicted before subjects are allocated. These types of subversion are defined according to these different causes as first-, second- and third-order selection bias, respectively [1].

Second-order selection bias may thus occur when methods for concealing the random allocation sequence are inadequate or when the random allocation sequence is wilfully unmasked, even when concealment has followed methods considered theoretically adequate. A systematic review conducted by the Cochrane Collaboration found that RCTs with inadequate or unknown allocation concealment might yield larger effect estimates than RCTs where concealment of the random allocation sequence was judged to be adequate, thus indicating a risk of effect size inflation due to second-order selection bias [3].

For prevention of second-order selection bias in RCTs the Cochrane Collaboration recommends the following: "Central allocation (including telephone-, web-based and pharmacy-controlled randomisation); sequentially numbered drug containers of identical appearance; sequentially numbered, opaque, sealed envelopes" [4]. The Cochrane Collaboration considers implementation of these allocation concealment measures to be adequate and, through its "risk of bias" (ROB) assessment tool, judges RCTs that report their use in allocation concealment as ensuring "low bias risk" [4]. However, the use of Cochrane's ROB tool for the assessment of allocation concealment adequacy in RCTs has been found to be of limited reliability, owing to low agreement between individual trial reviewers: $\kappa = 0.54$ (95% confidence interval [CI]: 0.29–0.79) [5], $\kappa = 0.50$ (95% CI: 0.36–0.63) [6] and $\kappa = 0.24$ (95% CI: 0.05–0.43) [7], as well as low agreement across reviewer pairs: $\kappa = 0.37$ (95% CI: 0.19–0.55) [7]. Although the assessment tool has been revised since these measures were taken, such revision extends only to a clarification that the domain "allocation concealment" falls into the domain "selection bias" [8]. Furthermore, the Cochrane approach for judging selection bias risk is based on the reported attempt by trialists to apply allocation concealment within RCTs but does not consider whether such attempt, even if judged as having included adequate methodology, was indeed successful.

In contrast, Berger has proposed the routine use of statistical testing within RCT methodology in order to establish whether the attempt of allocation concealment was unsuccessful and whether the effect of selection bias could not be averted by the end of the trial [1]. Such testing utilises statistical correlation of intervention outcomes with the propensity of trial subjects to be allocated to an experimental intervention group. The propensity of allocation of a trial subject to a certain type of intervention within an RCT is expressed through the reverse propensity score (RPS): the RPS reflects the propensity of one patient being allocated to one particular intervention group [1].

RPS-based testing forms the basis of the Berger–Exner test that was designed for the detection of third-order selection bias in RCTs [1]. Ideally, trialists themselves within their RCTs should apply such testing, by computing the RPS for each subject and using regression analysis to assess the association between RPS and intervention outcome per group. Subsequently, the RPS and intervention outcome per subject, as well as the regression results, should be included in the final trial report. Positive test results, indicating risk of selection bias, should be included in the discussion of presented effect estimates within the trial report.

While the RPS-based Berger–Exner test has been designed to detect the effect of third-order selection bias in RCTs, it was not designed for the detection of second-order selection bias effect. It was the aim of this study to adopt RPS-based testing accordingly and to establish, through use of RCT simulation, the potential accuracy of such testing for the detection of second-order selection bias effect in RCTs.

10.2 Materials and Methods

RCT simulations with varying parameters, as well as one biased and one non-biased trial scenario, were developed. In each trial scenario, RPS-based testing was conducted and test accuracy from the number of true positive (TP), true negative (TN), false positive (FP) and false negative (FN) test results were established.

10.2.1 Trial Simulation and Parameters

RCT simulations were developed by assuming two compared interventions groups (A and B), both with dichotomous outcomes being either intervention failure or success ($Y = 0$ or 1). RCT simulations consisted of (i) subject ID number sequence, (ii) RPS for each subject ID sequence representing the propensity of each subject to be allocated to Group A and (iii) dichotomous outcomes sequence (either $Y = 1$ or 0 per subject ID). Calculation of the RPS is best explained on the example of the propensity of one subject to be allocated to treatment group "A" along the concealed random sequence B-A-B-A (in the case of fixed block randomisation with block size 4 as randomisation method as example). Because the B-A-B-A sequence allows for two allocations into group A and two into group B, the RPS for the first allocation is $2:4 = 0.5000$. Once the first subject is allocated, the propensity of the next subject to be allocated to group A along the remaining A-B-A sequence is $2:3$. The resulting RPS = 0.6666 is higher than the previous one because now the A-B-A sequence allows for two allocations into group A but only one into group B. In line with this principle, the RPS for allocation into group A for the third and fourth subject is 1:2 = 0.5000 and 1:1 = 1.000, respectively.

The actual intervention effect in both group A and group B was of no interest and thus was set at zero ($Y = 0$). With an assumed 100% intervention failure rate for both groups, any comparison of both groups would yield no superior result of one above the other (= odds ratio [OR] 1.00). The assumption of such absolute treatment effect equivalence was further important as guarantor that any observed test result was due to the presence/absence of bias alone. Because this was the first study in this field with its outcome being unknown prior, the exclusion of treatment effects in both groups assisted in avoiding any ambiguity in the interpretation of results as whether RPS-based testing is able to detect, in principle, the effect of second-order selection bias or not.

Three parameter types—(i) subject number per group (N), (ii) trait frequency (TF) and (iii) type of randomisation method (RM)—were included into the simulation. A percentage of all subjects were assumed to contain a certain confounding characteristic/trait "X" that would cause success ($Y = 1$) with either intervention. The percentage of subjects with trait "X" in each trial was called the "trait frequency" (TF) and set at 10%, 20% or 50% of the total number of subjects (N), which in turn was set at 120, 240 or 480 subjects in each simulation group. Seven different randomisation methods were applied: fixed block randomisation with block size 4, 6 or 8; block randomisation with varying block sizes of 4, 6, 8 and with equal probability (1/3); as well as the maximal procedure [2] with a maximum tolerated imbalance (MTI) of 2, 3 or 4. From these parameters, a total of 63 different parameter sets for the RCT simulations were produced. All simulated RCTs were generated in four steps by use of R statistical software with the variables ID, BLOCK, TRT, RPS and Y (Appendix file 1).

10.2.2 Trial Scenarios

Without introduction of bias, neither intervention would lead to a superior outcome above the other. The trait "X" ($Y = 1$) subject distribution in both groups A and B was thus used as the bias indicator. Following a strict subject allocation according to random sequence in Scenario 1, all subjects with trait "X" were evenly distributed between both groups, thus resulting in an equal intervention effect in both groups. In Scenario 2, direct observation of the random sequence and knowing which of the subjects carried trait "X" formed the basis for allocating these subjects to group A, only. In this scenario the outcome of intervention A appeared superior to that of B, solely due to the allocation of trait X subjects in group A.

Scenario 1 was simulated by assigning the first (N*TF)/100% of subjects along the generated random allocation sequence (consisting of allocations to either group in a random manner) to $Y = 1$, and all remaining subjects of either group to $Y = 0$. Scenario 2 was simulated by assigning the first

(N*TF)/100% of subjects of group A (TRT = 1) within the generated random allocation sequence to $Y = 1$ and all other subjects of either group to $Y = 0$. A total of 25 RCT simulations ("runs") were conducted per parameter set.

For each parameter setting, the effect size inflation due to bias was computed as OR (with 95% CI), using RevMan 4.2.10 statistical software, and the results are presented in one forest plot for Scenarios 1 and 2.

10.2.3 Bias Testing

Testing was based on logistic regression analysis. The RPS for each subject ID sequence and the dichotomous outcomes ($Y = 1$ or 0) sequence per subject ID were entered into R statistical software programme for analysis. The RPS was considered as the independent variable, and the Y values were considered as the dependent variable. In contrast to the Berger–Exner test [1], regression analysis was not conducted for each intervention group separately, but for all subjects (N) of both intervention groups combined. The reason for such deviation was the difficulty in using logistic regression analysis if the entire cohort of one intervention group would have been formed by subjects with trait X ($Y = 1$) in Scenario 2. The p value was recorded and alpha was set at 1%, 5% or 20% as the threshold for a negative or positive test outcome.

A TN test outcome was recorded when the resulting p value was larger than 0.01, 0.05 or 0.20 for an alpha setting of 1%, 5% or 20%, respectively, in Scenario 1. A TP test outcome was recorded when the resulting p value was less than 0.01, 0.05 or 0.20 for alpha 1%, 5% or 20%, respectively, in Scenario 2. The number of FP results was established by subtracting the total number of TN test outcomes from the total number of runs per parameter set (Scenario 1). The number of FN results was established by subtracting the total number of TP results from the total number of runs per parameter set (Scenario 2).

For this test, the RPS-based mechanism of bias detection is explained on the basis of the allocation of trait X subjects (with positive treatment outcome, $Y = 1$) to RPS values (which can be large or small) as follows: When trait X subjects are evenly allocated, by strictly following the random allocation sequence, among both intervention groups in Scenario 1, the possibility for these to be evenly allocated to high and low RPS values is high. For that reason, it will often not be possible to detect a direct association beyond the play of chance between RPS values and treatment outcome ($Y = 0/1$) using logistic regression analysis and thus a TN test result is generated. Any observed FP results would be due to the association of some subjects to higher RPS values by chance.

In Scenario 2, allocation of trait X subjects is restricted to treatment group A, only, due to the subversive effect of second-order selection bias. Because the number of available RPS values in one intervention group is limited, the possibility is higher that trait X subjects are allocated to more high RPS values than in Scenario 1. In this case, logistic regression has a greater chance to detect a direct association between RPS values and treatment outcome and this would correctly generate TP test results. Any few observed FN results would again be due to the association of some subjects with positive treatment outcome ($Y = 1$) with lower RPS values by chance.

10.2.4 Data Analysis and Summary Measures

The total TN/FP and TP/FN results, for each alpha level, were entered into Meta-DiSc Version 1.4 statistical software [9] and the pooled test specificity and sensitivity (with 95% CI) were computed. Symmetric summary receiver operating characteristic (SROC) curves were generated using Meta-DiSc software for each alpha level. SROC curves give a graphical representation of the accuracy of predictions according to the established TP results-based sensitivity (y axis) and the FP results-based 1 − specificity (x axis). Poor test accuracy is indicated by SROC lines close to the rising diagonal and high test accuracy is indicated by steep rising SROC lines that pass close to the top left-hand corner of the graph [10].

Pooled diagnostic odds ratios (DOR, 95% CI) were used to investigate the influence of parameter N, TF and RM by computing them from the relevant TN/FP and TP/FN outcomes (Table 10.1) per alpha level. The DOR summarises the test sensitivity together with the test specificity into one single measure. It is defined as (TP × TN)/(FP × FN) [10]. A DOR value of (or close to) 1.00 suggests no predictive value of the test and corresponds to the rising diagonal in the SROC graphs. The higher the DOR value, the better the test accuracy [11].

All data pooling in this study followed the Der Simonian Laird random-effects model.

10.3 Results

In this study, second-order selection bias generated a deflection of the true effect estimate (OR 1.00) to point estimates falsely ranging from OR 0.00 to 0.04 (Scenario 2, Figure 10.1).

For each scenario, 1575 runs were conducted. In Scenario 1, the investigation yielded 1439, 1265 and 937 TN results for alpha 1%, 5% and 20%, respectively. In Scenario 2, the investigation yielded 1335, 1442 and 1536 TP results for alpha 1%, 5% and 20%, respectively (Appendix file 2).

10.3.1 Test Sensitivity and Specificity

A pooled sensitivity of 0.85 (95% CI: 0.83–0.86), 0.92 (95% CI: 0.90–0.93) and 0.98 (95% CI: 0.97–0.98) for alpha 1%, 5% and 20% was established,

TABLE 10.1

Generated Parameter Sets for Both Scenarios

PSN	RM	N	TF/n	PSN	RM	N	TF/n
01	Fixed/BS = 4	120	10/12	33	Varying	240	50/120
02			20/24	34		480	10/48
03			50/60	35			20/96
04		240	10/24	36			50/240
05			20/48	37	Maximal	120	10/12
06			50/120	38	procedure/		20/24
07		480	10/48	39	MTI = 2		50/60
08			20/96	40		240	10/24
09			50/240	41			20/48
10	Fixed/BS = 6	120	10/12	42			50/120
11			20/24	43		480	10/48
12			50/60	44			20/96
13		240	10/24	45			50/240
14			20/48	46	Maximal	120	10/12
15			50/120	47	procedure/		20/24
16		480	10/48	48	MTI = 3		50/60
17			20/96	49		240	10/24
18			50/240	50			20/48
19	Fixed/BS = 8	120	10/12	51			50/120
20			20/24	52		480	10/48
21			50/60	53			20/96
22		240	10/24	54			50/240
23			20/48	55	Maximal	120	10/12
24			50/120	56	procedure/		20/24
25		480	10/48	57	MTI = 4		50/60
26			20/96	58		240	10/24
27			50/240	59			20/48
28	Varying	120	10/12	60			50/120
29			20/24	61		480	10/48
30			50/60	62			20/96
31		240	10/24	63			50/240
32			20/48				

Note: PSN, parameter set number; RM, randomisation method; TF, trait frequency; N, subject number per group; n, number of subjects per TF; BS, block size; MTI, maximum tolerated imbalance; Fixed, fixed block randomisation; Varying, block randomisation with randomly varying block size 4, 6, 8 with equal probability (1/3).

respectively. A pooled test specificity of 0.91 (95% CI: 0.90–0.93), 0.80 (95% CI: 0.78–0.82) and 0.60 (95% CI: 0.57–0.62) for alpha 1%, 5% and 20% was established, respectively.

All SROC curves indicated the highest overall test accuracy when alpha was set at 1%.

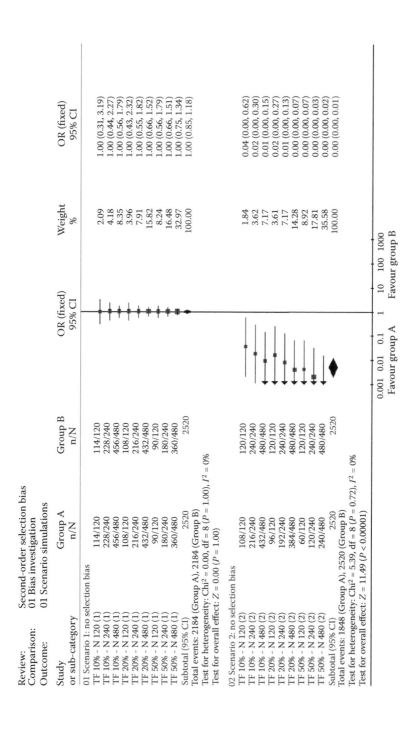

FIGURE 10.1

Inflation of effect size due to second-order selection bias. N, number of evaluated subjects; n, number of subjects with intervention failure ($Y = 0$); OR, odds ratio; CI, confidence interval; TF, trait frequency.

TABLE 10.2

Pooled DOR per Study Parameter

Study Parameter	Parameter Setting	Alpha 1%		Alpha 5%		Alpha 20%	
		DOR	95% CI	DOR	95% CI	DOR	95% CI
N	120	33	14–79	20	10–39	18	10–30
	240	116	45–295	71	32–156	36	21–61
	480	226	122–420	150	81–278	80	42–150
TF	10%	23	11–48	16	9–29	16	9–26
	20%	115	47–281	76	36–159	57	32–102
	50%	598	295–1212	233	122–448	82	44–155
RM	Fixed/BS=4	214	72–636	74	21–254	57	24–132
	Fixed/BS=6	74	16–347	52	12–221	37	12–108
	Fixed/BS=8	37	12–114	29	10–91	18	8–42
	Varying	115	28–470	47	16–136	22	11–46
	MP/MTI = 2	407	90–1845	136	58–318	84	33–214
	MP/MTI = 3	124	29–539	85	22–320	49	14–171
	MP/MTI = 4	61	12–324	40	11–150	28	12–64

Note: CI, confidence interval; Fixed, fixed block randomisation; BS, block size; MTI, maximum tolerated imbalance; Fixed, fixed block randomisation; Varying, block randomisation with randomly varying block size 4, 6, 8 with equal probability (1/3); MP, maximal procedure; MTI, maximum tolerated imbalance; N, subject number per group; TF, trait frequency; RM, randomisation method; (All decimals rounded) into the legend.

10.3.2 Trial Parameters and Test Accuracy

The results of the computed DOR values are shown in Table 10.2. Higher accuracy was observed at alpha 1%. The maximal procedure [2] with MTI = 2 was the randomisation method associated with the highest test accuracy (DOR 407; 95% CI: 90–1845). A higher accuracy with this procedure (in comparison to all other randomisation methods) was also observed at alpha 5% and 20% (Table 10.2). In general, test accuracy was associated with higher subject number (N), higher trait frequency (TF) and lower block size (BS) when fixed randomisation was used for all set alpha levels.

10.4 Discussion

The aim of this study was to establish, by use of RCT simulation, the potential accuracy of RPS-based testing for the detection of second-order selection bias effect in RCTs.

10.4.1 Study Method Limitations

The simulation method in this study only included one confounding factor, "trait X," which can be considered as an oversimplification. In reality, many confounding factors that are unknown may exist. Such factors may also affect each other in an enhancing or suppressive way and their combined influence on the study results may cause an over- or underestimation or may have no impact at all. The simulated "trait frequency" may also not be always static, but of varying value. Furthermore, the relationship of sample size to observed effect estimates in real-world RCTs might affect test accuracy.

In this simulation, an absolute equivalence in treatment effect between both interventions was assumed (= OR 1.00). Because this was the first study to investigate this topic and the accuracy of RPS testing for second-order selection bias effect was unknown, the assumption of absolute equivalence was important in order to avoid ambiguity in the interpretation of study results. However, even if interventions do have in reality similar treatment effects, such effects would seldom be absolute. Because the existence of treatment effects in intervention groups would theoretically affect the mechanism of the RPS test, the application of such testing in practice faces a conceptual dilemma: (i) a positive test result may either indicate presence of bias or true treatment effect of one intervention above that of the other in absence of bias; (ii) a negative test result may either indicate absence of bias or a true treatment effect of one intervention above that of the other but which has been masked by the second-order selection bias effect. On the other hand, such dilemma may exist only in theory. Bias will inflate the true treatment effect size in comparison to non-biased scenarios. Such inflation may be accompanied by distinct RPS patterns in the subject cohort with $Y = 1$ outcomes, which, in turn, may be affected by values of variables that were established in this study as of potential influence (N, TF, randomisation method) and even by other variables that still remain unknown. Also, the accuracy of bias effect detection may depend on the set alpha value as cutoff point. Any subsequent judgement of test accuracy will depend on the established TP, FN and TN, FP values, expressed as test sensitivity, specificity or DOR. As these assumptions are outside the scope of the current study, their corroboration may only be achieved through future research. However, the result of the current simulation study may have provided a basis and direction for such further investigations.

10.4.2 Discussion of Test Results

The results of this study suggest a high test sensitivity and specificity under conditions of absolute equivalence in treatment effect between compared intervention groups (OR = 1.00). The results were influenced by the chosen parameters and indicate higher test accuracy with lower alpha level (Figures 10.2 through 10.4), lower block size/MTI, higher sample size (N) and trait frequency (TF) (Table 10.2). The latter observation suggests that the higher the level of subversion, and thus the higher the inflating impact on the RCT effect estimate, the higher is the chance of detecting such subversion.

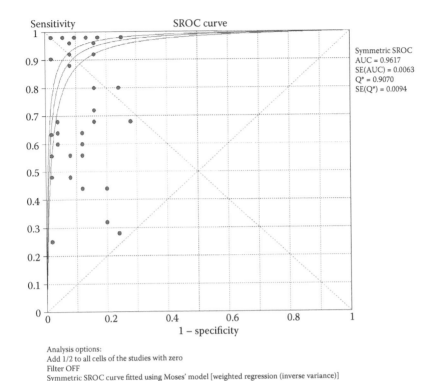

Analysis options:
Add 1/2 to all cells of the studies with zero
Filter OFF
Symmetric SROC curve fitted using Moses' model [weighted regression (inverse variance)]
Defined relevant region: All ROC spac

FIGURE 10.2
SROC curve concerning test accuracy at alpha level 1%. AUC, area under the curve; SE, standard error. Lines indicate the DOR with upper and lower confidence intervals.

The statistical detection of second-order selection bias effect in RCTs through RPS-based testing may indicate whether trial random allocation concealment was effective or not. The ROB assessment tool developed by the Cochrane Collaboration can currently not provide such information [4]. The detailed reporting of adequate concealment methods in RCT reports cannot provide any proof whether second-order selection bias was absent during the trials. It has been shown that RCTs with adequate concealment methods have lower effect estimates than those with inadequate or unknown methods [3]. However, such observations remain limited, because any prediction of effect size direction and magnitude changes cannot be made [3]. This may be due to the fact that the random allocation sequence of some RCTs with adequate concealment method was unmasked during the trial, and some RCTs that, although judged as of unknown concealment, were in fact adequately concealed.

The inclusion of an RPS-based selection bias test into routine trial methodology by those who conduct the trial, as well as the reporting of the

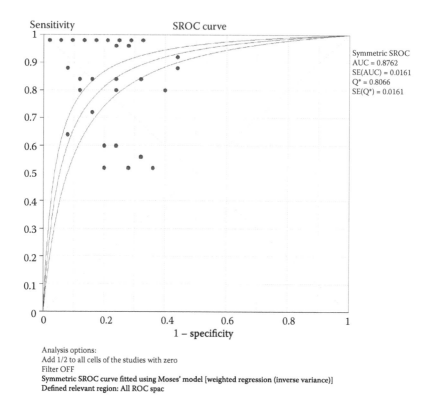

FIGURE 10.3
SROC curve concerning test accuracy at alpha level 5%. AUC, area under the curve; SE, standard error. Lines indicate the DOR with upper and lower confidence intervals.

dichotomous intervention outcome, the RPS score per trial subject and the test result may assist in the identification of RCTs with inadequate allocation concealment. Such information may assist in statistical bias correction [12,13] and in RCT evidence rating, as, for example, based on the proposed GRADE system [14].

Berger has argued for the inclusion of the Berger–Exner test for the detection of third-order selection bias effect into routine RCT methodology [1]. Third-order selection bias effect is caused in situations where allocation concealment is adequate and no direct observation of the random allocation sequence is possible. Subsequently, the random sequence is predicted on the basis of information regarding type and details of the applied randomisation method together with knowledge concerning choices for the allocation of previous subjects into different intervention groups. Investigation of the Berger–Exner test's accuracy to detect third-order selection bias effect indicates high sensitivity and specificity, particularly when the maximal procedure was

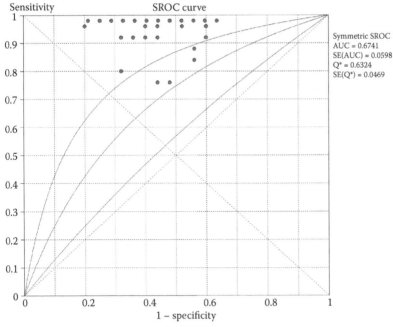

FIGURE 10.4
SROC curve concerning test accuracy at alpha level 20%. AUC, area under the curve; SE, standard error. Lines indicate the DOR with upper and lower confidence intervals.

used at MTI = 2 and with alpha set at 1% [15]. This study established the same for RPS-based second-order selection bias effect testing. However, unlike the Berger–Exner test, the accuracy of RPS-based second-order selection bias effect testing increased with higher subject number (N) and higher trait frequency (TF).

In this study, it was noted that the Berger–Exner test is unsuitable for second-order selection bias effect testing. However, RPS-based testing, following the method outlined in this study, would be inferior to the Berger–Exner test in the detection of third-order selection bias. For the latter, the application of regression analysis to the different intervention groups, separately, is needed in order to avoid false-positive results, especially when interventions with treatment effects are compared in RCTs [1]. Both tests cannot be applied *post hoc*, for example, during third-part appraisals in the form of systematic reviews of trials; thus, both require application and transparent reporting of its results by the trialists themselves.

The application of these tests in RCTs would ideally assure readers of trial reports that second- and third-order selection bias risk has been investigated and whether the validity of the presented RCT results were limited by the risk of such bias (or not).

Notwithstanding, in praxis, the existence of third-order selection bias will lead to a positive association between the RPS and intervention outcome, consequently generating a positive Berger–Exner test. A negative Berger–Exner test will indicate lack of such association and thus an absence of third-order selection bias. However, the Berger–Exner test will not be able to detect second-order selection bias effect since this type of selection bias does not rely on an association between RPS and intervention outcome. For that reason, a negative Berger–Exner test may be obtained either in the absence of any selection bias or in the presence of second-order selection bias effect. The latter may be successfully detected using the RPS-based testing method applied in this study, but only under condition of absolute equivalence in treatment effect between compared intervention groups (OR = 1.00). Unfortunately, because such absolute equivalence is unlikely in reality, such test may (pending further research) not be able to distinguish between second-order selection bias effect and genuine treatment effects.

For that reason, absence of evidence for bias, as shown by a negative Berger–Exner test result, cannot serve as evidence for selection bias absence.

10.4.3 Recommendation for Further Research

In this study, the application of RPS-based testing for second-order selection bias effect was investigated only for RCTs with dichotomous (or binary) outcomes. The presented results provide no guidance for bias testing in RCTs with continuous outcomes. Thus, further research in this field should focus on the adaptation of RPS-based testing for trials that measure effect sizes outside an intervention success/failure ($Y = 0$ or 1) dichotomy.

In addition, the questions as to why the observed test accuracy was higher for alpha level 1%, the maximal procedure and the reason for reduced test accuracy with higher block size/MTI require further investigation. An even more important question is whether the test remains accurate when interventions with variable treatment effects are compared, for example, in scenarios with true effect size differences between intervention groups and without being affected by bias. Here, further simulation studies need to establish the test's utility in differentiating between biased and non-biased scenarios under different parameter settings of sample size, trait frequency, alpha values or randomisation method. Further studies should also employ a larger number of runs per parameter set, in order to investigate the robustness of test results and a deeper investigation into the cause for FP/FN test results is warranted.

10.5 Conclusion

The results of this simulation-based investigation suggest that RPS-based testing can detect the effect of second-order selection bias with high sensitivity and specificity under conditions of absolute equivalence in treatment effect between the compared intervention groups. The accuracy appears highest when the maximal procedure as randomisation methods (with MTI = 2) is used and with alpha set at 1% as test threshold.

However, many questions remain as to the test's conceptual rationale and practicality and, thus, require further investigation before the test can be recommended for routine use.

Conflict of Interest Statement

The authors declare that the research was conducted in absence of any commercial or financial relationships that could be construed as a potential conflict of interest.

References

1. Berger VW. *Selection Bias and Covariate Imbalances in Randomised Clinical Trials.* Chichester, UK: John Wiley & Sons, Ltd. (2005).
2. Berger VW, Ivanova A, Knoll MD. Minimizing predictability while retaining balance through the use of less restrictive randomization procedures. *Stat Med* (2003) **22**(19):3017–28.
3. Odgaard-Jensen J, Vist GE, Timmer A, Kunz R, Akl EA, Schünemann H et al. Randomisation to protect against selection bias in healthcare trials. *Cochrane Database of Syst Rev* (2011) 4:MR000012.
4. Higgins JPT, Green S (editors). *Cochrane Handbook for Systematic Reviews of Interventions.* Version 5.1.0 [updated March 2011]. Table 8.5.d. The Cochrane Collaboration (2011). Available from www.cochrane-handbook.org.
5. Hartling L, Bond K, Vandermeer B, Seida J, Dryden DM, Rowe BH. Applying the risk of bias tool in a systematic review of combination long-acting beta-agonists and inhaled corticosteroids for persistent asthma. *PLOS One* (2011) 6:e17242.
6. Hartling L, Ospina M, Liang Y, Dryden DM, Hooton N, Krebs Seida J et al. Risk of bias versus quality assessment of randomised controlled trials: Cross sectional study. *BMJ* (2009) **339**:b4012.
7. Hartling L, Hamm MP, Milne A, Vandermeer B, Santaguida PL, Ansari M et al. Testing the risk of bias tool showed low reliability between individual

reviewers and across consensus assessments of reviewer pairs. *J Clin Epidemiol* (2012) **66**(9):973–81.

8. Higgins JPT, Green S (editors). *Cochrane Handbook for Systematic Reviews of Interventions.* Version 5.1.0 [updated March 2011]. Table 8.5.b. The Cochrane Collaboration (2011). Available from www.cochrane-handbook.org.

9. Zamora J, Abraira V, Muriel A, Khan K, Coomarasamy A. Meta-DiSc: A software for meta-analysis of test accuracy data. *BMC Med Res Methodol* (2006) **6**:31.

10. Deeks JJ. Systematic reviews of evaluations of diagnostic and screening tests. In: Egger M, Smith GD, Altman DG (editors): *Systematic Reviews in Health Care.* BMJ Publishing group (2001):248.

11. Glas AS, Lijmer JG, Prins MH, Bonsel GJ, Bossuyt PM. The diagnostic odds ratio: A single indicator of test performance. *J Clin Epidemiol* (2003) **56**(11):1129–35.

12. Berger VW. The reverse propensity score to detect selection bias and correct for baseline imbalances. *Stat Med* (2005) **24**(18):2777–87.

13. Ivanova A, Barrier RC Jr, Berger VW. Adjusting for observable selection bias in block randomized trials. *Stat Med* (2005) **24**(10):1537–46.

14. Guyatt GH, Oxman AD, Vist G, Kunz R, Brozek J, Alonso-Coello P et al. GRADE guidelines: 4. Rating the quality of evidence—Study limitations (risk of bias). *J Clin Epidemiol* (2011) **64**(4):407–15.

15. Mickenautsch S, Fu B, Gudehithlu S, Berger VW. Accuracy of the Berger–Exner test for detecting third-order selection bias in randomised controlled trials: A simulation-based investigation. *BMC Med Res Methodol* (2014) **14**:114.

Appendix File 1 Four-Step RCT Simulation with R code "rps.gen"

```
# Function to generate RPS, given treatment setting and block
setting
RPS.gen <- function(trt, # Treatment setting, require binary
here
block # Block setting, can be different block size
){
index <- unique(block)
rps <- NULL
for(i in 1:length(index)){
trt.sub <- trt[which(block==index[i])]
rps.sub <- sapply(1:length(trt.sub),
function(j){
ifelse(j==1, sum(trt.sub)/length(trt.sub),
sum(trt.sub[-(1:(j-1))])/length(trt.sub[-(1:(j-1))]))})
rps <- c(rps,rps.sub)
}
return(rps)
}
```

Appendix File 2 Test Results (Raw Data)

| | Alpha 1% | | | | Alpha 5% | | | | Alpha 20% | | | |
| | Scenario 1 | | Scenario 2 | | Scenario 1 | | Scenario 2 | | Scenario 1 | | Scenario 2 | |
PSN	TN	FP	TP	FN	TN	FP	TP	FN	TN	FP	TP	FN
1	25	0	6	19	20	5	13	12	15	10	23	2
2	24	1	16	9	22	3	20	5	16	9	25	0
3	24	1	25	0	22	3	25	0	17	8	25	0
4	25	0	16	9	22	3	21	4	20	5	24	1
5	25	0	25	0	21	4	25	0	16	9	25	0
6	24	1	25	0	21	4	25	0	15	10	25	0
7	23	2	25	0	19	6	25	0	14	11	25	0
8	22	3	25	0	22	3	25	0	17	8	25	0
9	24	1	25	0	22	3	25	0	15	10	25	0
10	20	5	8	17	16	9	13	12	14	11	19	6
11	25	0	14	11	19	6	21	4	11	14	25	0
12	22	3	25	0	19	6	25	0	17	8	25	0
13	22	3	16	9	21	4	18	7	16	9	23	2
14	25	0	23	2	25	0	25	0	19	6	25	0
15	23	2	25	0	18	7	25	0	15	10	25	0
16	21	4	23	2	19	6	25	0	15	10	25	0
17	22	3	25	0	20	5	25	0	16	9	25	0
18	22	3	25	0	20	5	25	0	17	8	25	0
19	25	0	12	13	23	2	16	9	17	8	20	5
20	21	4	17	8	14	11	22	3	12	13	24	1
21	24	1	25	0	22	3	25	0	11	14	25	0
22	20	5	11	14	17	8	14	11	11	14	21	4
23	19	6	20	5	18	7	24	1	13	12	25	0
24	19	6	25	0	17	8	25	0	12	13	25	0
25	21	4	20	5	14	11	23	2	9	16	25	0
26	21	4	24	1	17	8	25	0	11	14	25	0
27	24	1	25	0	22	3	25	0	20	5	25	0
28	23	2	12	13	20	5	15	10	11	14	22	3
29	22	3	14	11	19	6	20	5	17	8	23	2
30	21	4	25	0	19	6	25	0	13	12	25	0
31	24	1	15	10	21	4	21	4	14	11	23	2
32	24	1	25	0	21	4	25	0	18	7	25	0
33	24	1	25	0	21	4	25	0	13	12	25	0
34	22	3	25	0	18	7	25	0	14	11	25	0
35	22	3	25	0	19	6	25	0	13	12	25	0
36	25	0	25	0	20	5	25	0	15	10	25	0

(Continued)

| PSN | Alpha 1% | | | | Alpha 5% | | | | Alpha 20% | | | |
| | Scenario 1 | | Scenario 2 | | Scenario 1 | | Scenario 2 | | Scenario 1 | | Scenario 2 | |
	TN	FP	TP	FN	TN	FP	TP	FN	TN	FP	TP	FN
37	23	2	14	11	22	3	20	5	14	11	24	1
38	23	2	25	0	22	3	25	0	18	7	25	0
39	25	0	25	0	22	3	25	0	15	10	25	0
40	23	2	22	3	19	6	24	1	13	12	25	0
41	25	0	25	0	23	2	25	0	18	7	25	0
42	24	1	25	0	23	2	25	0	20	5	25	0
43	24	1	25	0	21	4	25	0	17	8	25	0
44	25	0	25	0	23	2	25	0	17	8	25	0
45	25	0	25	0	22	3	25	0	18	7	25	0
46	22	3	11	14	19	6	15	10	13	12	19	6
47	24	1	17	8	23	2	22	3	16	9	25	0
48	23	2	25	0	19	6	25	0	14	11	25	0
49	21	4	18	7	19	6	21	4	15	10	24	1
50	25	0	25	0	23	2	25	0	19	6	25	0
51	23	2	25	0	20	5	25	0	15	10	25	0
52	23	2	24	1	22	3	25	0	18	7	25	0
53	23	2	25	0	20	5	25	0	17	8	25	0
54	24	1	25	0	24	1	25	0	19	6	25	0
55	19	6	7	18	18	7	13	12	10	15	23	2
56	18	7	17	8	15	10	20	5	10	15	24	1
57	22	3	25	0	18	7	25	0	13	12	25	0
58	22	3	15	10	17	8	21	4	13	12	25	0
59	23	2	23	2	20	5	25	0	10	15	25	0
60	25	0	25	0	23	2	25	0	18	7	25	0
61	24	1	25	0	19	6	25	0	13	12	25	0
62	21	4	25	0	20	5	25	0	13	12	25	0
63	21	4	25	0	19	6	25	0	12	13	25	0
PSN				Parameter set number								
TN				True negative								
FP				False positive								
TP				True positive								
FN				False negative								

11

The Berger–Exner Test to Detect Third-Order Selection Bias in the Presence of a True Treatment Effect

Steffen Mickenautsch, Bo Fu, and Vance W. Berger

CONTENTS

11.1 Background...159
11.2 Methods..160
    11.2.1 Trial Simulation...161
    11.2.2 Study Scenarios and Parameters 162
    11.2.3 Bias Testing 163
    11.2.4 Data Analysis 164
11.3 Results 164
11.4 Discussion..165
    11.4.1 Limitations and Strengths of the Study Method 165
    11.4.2 Discussion of Results..167
    11.4.3 Recommendation for Further Research.....................................169
11.5 Conclusion 169
Authors' Contributions...170
Conflict of Interest..170
References..170

11.1 Background

Selection bias is a form of systematic error that interferes with the internal validity of clinical trials [1] and may lead to favouring one treatment group over another [2]. "Third-order residual" selection bias is present when, despite the application of an adequate allocation concealment process, subversion has occurred through correct prediction of the random sequence allocation [1]. The focus of this study is the situation in which this prediction is acted upon. In this case, the investigators may select healthier patients, or those patients more likely to respond well, when their treatment is up next, and sicker patients when the control is up next. The Berger–Exner test is the first analysis that aims to directly detect third-order selection bias by studying the mechanism through which it occurs. The test consists of a regression analysis, including the reverse propensity score (RPS) as the independent variable and the binary outcome as the dependent variable [1]. Regression analysis is conducted separately per intervention group and the resulting p values as test thresholds are utilised. However, the result of a simple literature search in PubMed and GoogleScholar (30 September 2013), using "Berger–Exner test" as the search term, suggests that, since its development, the test has been adopted as part of trial methodology in a few randomised controlled trials (RCTs) only [3]. Reasons for the apparent low utilisation of the Berger–Exner test in RCTs may be ascribed to a lack of information regarding its test accuracy. Against this background, a first trial simulation study was conducted [3]. During this previous study, simulated RCTs with different parameter settings and dichotomous outcomes were generated and subjected to bias-free and selection bias scenarios. The treatment effect of the trial's test and control groups was set at zero so that neither group yielded any superior results above the other. In addition, a "trait frequency" (TF) was defined as a percentage of subjects with a certain "Trait X" that would cause intervention success regardless of the type of intervention group. In a non-biased scenario, such trait was distributed evenly amongst both groups. In a scenario where third-order selection bias was simulated, the biased distribution of subjects with this trait lead to artificially higher effect estimates in one group above the other. Such simulated trial settings allowed for investigation of the bias effect alone. The Berger–Exner test was applied in both scenarios and the pooled test sensitivity and specificity for various alpha levels were computed.

The study result showed a test sensitivity of 1.00 and a test specificity of 0.94 at alpha level 1%, thus suggesting that the Berger–Exner test is highly accurate for detecting third-order selection bias.

However, the absolute equivalence of treatment effects does not reflect real trial conditions. Against this background, the aim of this study was to assess the accuracy of the Berger–Exner test in identifying "third-order residual selection bias" when effect sizes of competing interventions are not equal on the basis of relevant simulations for RCTs with measured dichotomous outcomes.

11.2 Methods

In order to investigate the accuracy of the Berger–Exner test for detecting third-order selection bias in RCTs and in line with previous study methods [3], the numbers of true positive (TP), true negative (TN), false positive (FP) and false negative (FN) results were observed from RCT simulations. A "true positive" test result was observed when the test indicated a positive result in the presence of bias and a "true negative" test result was observed when the test indicated a negative result in the absence of bias. Accordingly, a "false positive" test result was observed when the test indicated a positive result in the absence of third-order selection bias and a "false negative" test result was observed when the test indicated a negative result in the presence of third-order selection bias. These numbers were converted into the following outcome measures: the test sensitivity (SS), specificity (SP), diagnostic odds ratio (DOR), positive and negative predictive value (PPV/NPV) and positive and negative likelihood ratio (LR+/LR–).

The test sensitivity, also called the TP rate, is defined as the proportion of cases correctly identified with third-order selection bias (TP) in relation to all cases with third-order selection bias. The test specificity, also called the TN rate, is defined as the proportion of cases correctly identified without third-order selection bias (TN) in relation to the total number of cases without third-order selection bias [4]. For high accuracy, it is ideal that sensitivity and specificity have somewhat reasonably high percentages, that is, >80%. In order to obtain a summarised predictive value of the two rates, the DOR was computed. The DOR combines sensitivity and specificity into one single predictive summary measure and is defined as DOR = (TP × TN)/(FP × FN) [4]. The DOR may range from zero to infinity, has no pre-defined cutoff threshold and is utilised for comparing the predictive evidence strength of different parameter settings.

The PPV and NPV are defined as the proportion of cases with third-order selection bias in relation to the total number of positive test results and the proportion of cases without third-order selection bias in relation to the total number of negative test results, respectively. In this study with an equal number of biased and unbiased cases, the LR+ value represents the ratio of TP results to FP results and the LR– value represents the ratio of FN results to TN results. A threshold of LR+ > 10 and LR– < 0.1 are indicative for a highly accurate test [4].

Based on previous observations that indicated highest test accuracy under trial simulation without assumed effect sizes for test and control interventions [3], the following simulation settings were chosen: maximal procedure with the maximum tolerated imbalance (MTI) [5] set at 2; sample size $N = 500$ per intervention group; and alpha for statistical significance at 1% and 10%. A TF was defined as the percentage of subjects with a characteristic/trait "X". Such trait was assumed to function as a confounding factor that would cause intervention success ($Y = 1$), regardless of the intervention group, and was set at 20% of the sample size per intervention group.

11.2.1 Trial Simulation

RCT simulations were obtained by assuming the comparison of two interventions (Intervention A and B) with dichotomous outcomes (Intervention failure: $Y = 0$; Intervention success: $Y = 1$). A simulated RCT consisted of three components:

    i. A sequence of subject ID (accession) numbers

    ii. A sequence of the RPS [6] per subject ID with regard to the propensity of the subject to be allocated to Group A

    iii. A sequence of dichotomous outcomes per subject ID ($Y = 1$ or 0)

The RPS reflects the probability of allocation of a patient to a particular intervention group [1]. For example, with block size 4 and an [ABAB] block, the sequence of RPS values for allocation to group A would be 2/4, 1/3, 1/2, 0/1, respectively, reflecting the ratio of the number of remaining A allocations within the block to overall remaining allocations within the block.

11.2.2 Study Scenarios and Parameters

Two scenarios were simulated: without third-order selection bias (Scenario 1) and with third-order selection bias (Scenario 2). When subject allocation strictly followed a true random sequence, all subjects with trait "X" were evenly distributed between groups A and B (= Scenario 1). Subversion of the random allocation by correct prediction of the random sequence through use of the RPS and knowledge about which of the subjects carry trait "X" allowed allocation of these subjects in favour of a particular intervention group (= Scenario 2).

Scenario 1 was simulated, using R statistical software, by assigning the first (N*TF)/100% of participants to $Y = 1$, and all others to $Y = 0$. Scenario 2 was simulated, using R statistical software, by assigning $Y = 1$ in accordance with the highest RPS. For each parameter set, 1000 individual random sequences ("runs") were generated.

In addition, two types of parameter sets were generated:

    i. Treatment B being a placebo (without treatment effect) and treatment A being an effective treatment with a treatment effect being set to be either 0%, 5%, 10% and 20% above that of treatment B.

    ii. Treatment B being also an effective treatment with 5%, 10% and 15% of all subjects getting better. However, treatment A was set to always have a 5% higher treatment effect than B.

A total of 14 different parameter sets were generated and are presented in Table 11.1. Parameter sets 08–14 were identical to that of parameter sets 01–07, with the differences that the former were biased in favour of group B while the latter were biased in favour of group A.

TABLE 11.1

Generated Parameter Sets

Parameter Set/Nr.	N (per Group)	TF	TE		Biased in Favour of	Alpha
			Group A	Group B		
01	500	20%	0%	0%	Group A	1% and 10%
02	500	20%	5%	0%	Group A	1% and 10%
03	500	20%	10%	0%	Group A	1% and 10%
04	500	20%	20%	0%	Group A	1% and 10%
05	500	20%	10%	5%	Group A	1% and 10%
06	500	20%	15%	10%	Group A	1% and 10%
07	500	20%	20%	15%	Group A	1% and 10%
08	500	20%	0%	0%	Group B	1% and 10%
09	500	20%	5%	0%	Group B	1% and 10%
10	500	20%	10%	0%	Group B	1% and 10%
11	500	20%	20%	0%	Group B	1% and 10%
12	500	20%	10%	5%	Group B	1% and 10%
13	500	20%	15%	10%	Group B	1% and 10%
14	500	20%	20%	15%	Group B	1% and 10%

Note: N, number of trial subjects; TF, trait frequency; TE, treatment effect.

All RCT simulations were conducted in four steps, using R statistical software, based on the generated variables: ID, BLOCK, TRT, RPS, Y.

- Step 1: Generation of subject identification (ID) $i = 1{:}N$.
- Step 2: Generation of randomisation blocks/MTI (BLOCK) and randomisation.
- Step 3: Generation of two-arm treatment (TRT) in each block. The maximal procedure (MP), the generation of treatment sequence, was based on the MTI [5]; that is, the qualified random sequence should satisfy $I(D) < = MTI$ and $I_N(D) = 0$, where $I_k(D) = |S_{k,A}(D) - S_{k,B}(D)|$, and $S_{k,A}(D) = sum(X_i(D))$, $I = 1,...,k$, and $S_{k,B}(D) = k - S_{k,A}(D)$, where $X_i(D) = 1$, if sequence D assigns the ith patient to treatment group A, $X_i(D) = 0$, if sequence D assigns the ith patient to treatment group B.
- Step 4: Calculation of the RPS from the generated block and treatment, using code "rps.gen" in R (Additional file 1) on the basis of the block information (BLOCK) and the treatment information (TRT).

11.2.3 Bias Testing

The Berger–Exner test was applied for bias testing. In order to investigate the influence of various alpha levels on the test accuracy, alpha was set at 1% and 10%.

A TN result was established when both p values for intervention groups A and B were above 0.01 and 0.10 (two sided). A TP result was established when at least one of the p values of either intervention group was below 0.01 and 0.10. The number of FP results was calculated by subtracting the total number of TN from the total number of runs per parameter set; that is, FP = 1000 − TN. The number of FN results was calculated by subtracting the total number of TP from the total number of runs per parameter set; that is, FN = 1000 − TP.

11.2.4 Data Analysis

From the 14 separate parameter sets, the established total numbers of TN/FP and TP/FN results were used to compute all outcome measures.

The influence of biasing the different intervention groups and the difference in treatment effect between the groups was investigated. To investigate whether biasing the different intervention groups had any influence on the outcomes, the SP, PPV and LR+ values of parameter set 01–07 from testing at alpha 1% level were statistically compared with that of parameter sets 08–14 using the Mann–Whitney U test. Linear regression was conducted in order to investigate whether differences in treatment effect between the groups had any influence on the outcomes. The effect size differences were used as the independent variable and the SP, PPV and LR+ values were used as the dependent variables. Critical value (alpha) was set at 5% for all investigations.

11.3 Results

In total, 1000 simulation runs were conducted for each of the 14 parameter sets per alpha level. From these, 13,503 and 10,901 yielded TN results for alpha 1% and 10%, respectively, and 14,000 yielded TP results for alpha 1% and 10%, each (Table 11.2).

The results of all computed outcome measures are presented in Table 11.3. The sensitivity (SS) and NPV were 1.00, the LR− was zero and the DOR was infinite for all parameter sets and set alpha levels.

The specificity (SP) ranged between 0.74 and 0.81 for alpha level 10% and between 0.95 and 0.98 for alpha level 1%. The PPV ranged between 0.79 and 0.84 for alpha level 10% and between 0.95 and 0.98 for alpha level 1%. The LR+ ranged from 3.86 to 5.32 for alpha level 10% and from 21.28 to 40.00 for alpha level 1%.

Neither biasing of the different intervention groups nor the differences in treatment effect between the groups appeared to have had any effect on the test accuracy (Table 11.4).

TABLE 11.2

Summary of True Negative/Positive Results

Observed Outcomes	Parameter Set/Nr.						
	01	02	03	04	05	06	07
Alpha = 1%							
TN	967	966	962	968	957	965	965
TP	1000	1000	1000	1000	1000	1000	1000
Alpha = 10%							
TN	812	793	774	771	787	789	785
TP	1000	1000	1000	1000	1000	1000	1000
	08	09	10	11	12	13	14
Alpha = 1%							
TN	969	961	966	953	963	975	966
TP	1000	1000	1000	1000	1000	1000	1000
Alpha = 10%							
TN	795	773	762	781	741	776	762
TP	1000	1000	1000	1000	1000	1000	1000

Note: TN, true-negative results; TP, true-positive results.

11.4 Discussion

The aim of this study was to assess the accuracy of the Berger–Exner test in identifying "third-order residual selection bias" when effect sizes of competing interventions are not equal on the basis of relevant simulations for RCTs with measured dichotomous outcomes.

11.4.1 Limitations and Strengths of the Study Method

In this simulation study, "trait X" as only one single confounding factor was included. This may be an oversimplification because many unknown confounding factors with effect on trial outcomes may exist. These may interact in either an enhancing or suppressive form. Their combined influence may also lead to over- or underestimation or may have no impact at all. The TF may also not always be static, but may vary in value.

In addition, as the study simulation relied on binary values ($Y = 0/Y = 1$) as the dependent variable, difficulties in using logistic regression were observed. For scenario 2 (third-order selection bias), it may be the case that for all subjects in one treatment group, there may be only one value of outcome ($Y = 0$ or $Y = 1$). In such cases, logistic regression fails with Y as the dependent variable. However, with linear regression, the RPS value can be

TABLE 11.3

Outcome Measures/Results

Parameter Set/Nr.	SS	SP	DOR	PPV	NPV	LR+	LR−
Outcome measures results (Alpha = 1%)							
01	1	0.967	Infinite	0.968	1	30.30	0
02	1	0.966	Infinite	0.967	1	29.41	0
03	1	0.962	Infinite	0.963	1	26.32	0
04	1	0.968	Infinite	0.968	1	31.25	0
05	1	0.957	Infinite	0.958	1	23.26	0
06	1	0.965	Infinite	0.966	1	28.57	0
07	1	0.965	Infinite	0.966	1	28.57	0
08	1	0.969	Infinite	0.970	1	32.26	0
09	1	0.961	Infinite	0.962	1	25.64	0
10	1	0.966	Infinite	0.967	1	29.41	0
11	1	0.953	Infinite	0.955	1	21.28	0
12	1	0.963	Infinite	0.964	1	27.02	0
13	1	0.975	Infinite	0.976	1	40.00	0
14	1	0.966	Infinite	0.967	1	29.41	0
Outcome measures results (Alpha = 10%)							
01	1	0.812	Infinite	0.842	1	5.32	0
02	1	0.793	Infinite	0.829	1	4.83	0
03	1	0.774	Infinite	0.816	1	4.42	0
04	1	0.771	Infinite	0.814	1	4.37	0
05	1	0.787	Infinite	0.824	1	4.69	0
06	1	0.789	Infinite	0.826	1	4.73	0
07	1	0.785	Infinite	0.823	1	4.65	0
08	1	0.795	Infinite	0.830	1	4.88	0
09	1	0.773	Infinite	0.815	1	4.41	0
10	1	0.762	Infinite	0.808	1	4.20	0
11	1	0.781	Infinite	0.820	1	4.57	0
12	1	0.741	Infinite	0.794	1	3.86	0
13	1	0.776	Infinite	0.817	1	4.46	0
14	1	0.762	Infinite	0.808	1	4.20	0

Note: DOR, diagnostic odds ratio; LR+, positive likelihood ratio; LR−, negative likelihood ratio; NPV, negative predictive value; PV, positive predictive value; SS, sensitivity; SP, specificity

utilised as the independent variable, which is why it was selected as the analytical method.

In the previous simulation study on the same topic, a range of 63 separate parameters were investigated [3]. The results showed that using the maximal procedure yields higher test accuracy than using block randomisation as the randomisation method was included in the trial simulation, that MTI = 2 yielded higher test accuracy than larger MTI settings and that a set alpha

TABLE 11.4

Influence of Parameter Settings on Test Accuracy for Alpha 1%

(1) Biasing of the Different Intervention Groups (Mann–Whitney U Test)				
Data	n_A	n_B	U	p Value
SP	7	7	23	0.85
PPV	7	7	23	0.85
LR+	7	7	23	0.85

(2) Differences in Treatment Effect between the Groups (Linear Regression Analysis)							
				Intercept			
Data	DS	Adj R^2	Coefficient	SE	t	p Value	95% CI
SP	14	0.07	−0.0003	0.0002	−1.42	0.18	−0.0008 to 0.0002
PPV	14	0.09	−0.0003	0.0002	−1.49	0.16	−0.0008 to 0.0002
LR+	14	0.02	−0.2247	0.1982	−1.13	0.28	−0.6565 to 0.2071

Note: CI, confidence interval; DS, number of data sets; LR+, positive likelihood ratio; nA, number of investigated units biased in favour of group A; nB, number of investigated units biased in favour of group B; PPV, positive predictive value; SE, standard error; SP, specificity

of 1% yielded higher test accuracy than an alpha set at 5% or 20%. For that reason, these settings were chosen for this study. An additional alpha level of 10% was also included in order to corroborate whether higher test accuracy was retained at 1% even when treatment effect differences were included in the trial simulation. Furthermore, the previous study employed only 25 runs per parameter set [3]. In this study, this was increased to 1000 runs per parameter set, thus greatly increasing the precision of the measured outcomes.

In addition, the trial simulation employed in the previous study was based on an assumed absolute equivalence (odds ratio = 1.00) in effect size between both competing interventions [3]. However, in reality, even if interventions with almost the same treatment effects were compared in clinical trials, such effects would seldom be absolutely equal. Therefore, a range of treatment effect differences between the competing interventions was implemented within this study (Table 11.1). These differences ranged from no differences (Parameter set/Nr. 01 and 08), an increasing effect in one intervention with no effect of the competing one (Parameter set/Nr. 02–04 and 09–11) to an increasing effect in both interventions but with remaining effect size differences (Parameter set/Nr. 05–07 and 12–14). In that way, a more realistic simulation of possible intervention outcomes was achieved.

11.4.2 Discussion of Results

The results of this study's trial simulation indicate high accuracy in all outcome measures (SS, SP, DOR, NPV, PPV, LR+ and LR−) of the Berger–Exner test to detect third-order selection bias in RCTs with dichotomous outcomes that compare interventions with different treatment effects (Table 11.3).

Throughout the trial simulations of all parameter settings and alpha levels, no FN results were observed. In theory, biased cases may be either correctly identified as biased, yielding a (true) positive test result (TP), or incorrectly identified as non-biased, yielding a (false) negative test result (FN). The Berger–Exner test correctly identified all biased cases (in Scenario 2), thus rendering its sensitivity (being the percentage of TP results from all biased cases) at a maximum of 100% (SS = 1.00). Consequently, the chance that the Berger–Exner test presented a biased case as (false) negative (expressed as the LR–) was zero. In addition, in the absence of any FN test results, all negative test results (100%) indicated correctly non-biased cases, which was expressed as NPV = 1.00 (Table 11.3).

For the same reason, the DOR (DOR = TP × TN/FP × FN) values for all parameter sets were considered to be infinite because the null FN value as denominator generated a division by zero. As the DOR values of all parameter sets were considered to be infinite, the DOR became unsuitable for comparing effects of different parameter settings on the accuracy of the Berger–Exner test.

The test sensitivity established in this study confirms previous findings (SS = 1.00; 95% CI: 0.99–1.00) for alpha 1% and the maximal procedure (MTI = 2) for trial randomisation [3].

Non-biased cases may be either correctly identified as non-biased, yielding a (true) negative test result (TN), or incorrectly identified as biased, yielding a (false) positive result (FP). To this regard, the Berger–Exner test correctly identified 74%–81% of all non-biased cases (in Scenario 1) at alpha 10% and 95%–98% of all cases at alpha 1% as negative. The test specificity can thus be considered as very high, albeit somewhat lower than its test sensitivity. The test specificity established in this study (SP range = 0.95–0.98) also confirms previous findings (SP = 0.94; 95% CI: 0.93–0.96) for alpha 1% and the maximal procedure (MTI = 2) for trial randomisation [3]. The observed minor differences in results between both studies may be ascribed to the higher precision in this study due to the larger number of runs per parameter set.

The chance (expressed as the LR+) of a biased case to yield a positive Berger–Exner test result was about 4–5 times higher than for a non-biased case at alpha 10% and about 21–40 times higher at alpha 1%. Furthermore, the established PPVs indicate that 79%–84% of all positive test results identified true biased cases (in Scenario 2) correctly at alpha 10% and 95%–98% at alpha 1%. The higher test accuracy at reduced alpha settings confirms previous observations [3].

Differences in further parameter settings, such as the treatment effect of compared interventions in the simulated trials, proved to have no effect on test accuracy. The test accuracy was furthermore not affected by the choice of which intervention group to benefit through third-order selection bias (Table 11.4).

As with previous simulation results [3], it remains unclear how these results relate to the available evidence provided by real-world RCTs that have utilised the Berger–Exner test. However, because of the current lack of RCTs that have used the Berger–Exner test [3], more in-depth considerations as to the practical meaning of the reported test results cannot yet be made.

Notwithstanding these limitations, a routine application of the Berger–Exner test in RCTs may aid in providing an empirical answer to the question as to whether concealment of the random allocation sequence was effective. RCT reports are unable (by virtue of mere stating in detail the use of adequate concealment methods) to provide proof that third-order selection bias is absent. Such lack of proof forms the basis for doubt whether some RCTs, which reported adequate concealment, are compromised by correct prediction, without any unmasking of the random allocation sequence. If the Berger–Exner test is routinely included in RCT methodology and the intervention outcome and RPS score per trial subject are reported, then the test result may aid in the correct distinction between RCTs with adequate or inadequate concealment. Such information may help in statistical RPS-based bias correction [6,7] and aid in the rating of RCT evidence quality. The GRADE system recommends quality downgrading of RCT evidence when adequate allocation concealment is lacking. However, empirical evidence in support of GRADE criteria is limited and the application of the GRADE system has been shown to provide inconsistent results [8].

11.4.3 Recommendation for Further Research

General adoption of the Berger–Exner test with alpha set at 1% into the RCT methodology of future studies that also use the maximal procedure with MTI = 2 for randomisation could aid the corroboration of the previous [3] and current simulation results. In addition, further investigations are required to explain why the test accuracy is reduced at higher alpha settings (i.e., >1%) and higher MTI. In addition, an investigation of the test accuracy for RCTs with continuous outcomes is needed.

11.5 Conclusion

The results of this simulation-based investigation suggest that in the presence of true treatment effects, the Berger–Exner test may successfully detect third-order selection bias at high accuracy under the condition that the maximal procedure with MTI = 2 as the randomisation method is used with alpha set at 1%.

Authors' Contributions

Steffen Mickenautsch developed the concept and outline of this study; Bo Fu developed the code "rps.gen" in R statistical software and conducted the statistical analysis. All authors contributed to the writing of the manuscript.

Conflict of Interest

The authors declare that they have no conflict of interest.

References

1. Berger VW, Exner DV: Detecting selection bias in randomized clinical trials. *Control Clin Trials* 1999, 20:319–327.
2. Gluud LL: Bias in clinical intervention research. *Am J Epidemiol* 2006, 163:493–501.
3. Mickenautsch S, Fu B, Gudehithlu S, Berger VW: Accuracy of the Berger–Exner test for detecting third-order selection bias in randomised controlled trials: A simulation-based investigation. *BMC Med Res Methodol* 2014, 14:114.
4. Deeks JJ: Systematic reviews of evaluations of diagnostic and screening tests. In: Egger M, Smith GD, Altman DG (editors): *Systematic Reviews in Health Care.* BMJ Publishing group, 2001: pp. 248.
5. Berger VW, Ivanova A, Knoll MD: Minimizing predictability while retaining balance through the use of less restrictive randomization procedures. *Stat Med* 2003, 22:3017–3028.
6. Berger VW: The reverse propensity score to detect selection bias and correct for baseline imbalances. *Stat Med* 2005, 24:2777–2787.
7. Ivanova A, Barrier RC Jr, Berger VW: Adjusting for observable selection bias in block randomized trials. *Stat Med* 2005, 24:1537–1546.
8. Guyatt GH, Oxman AD, Vist G, Kunz R, Brozek J, Alonso-Coello P, Montori V, Akl EA, Djulbegovic B, Falck-Ytter Y, Norris SL, Williams JW Jr, Atkins D, Meerpohl J, Schünemann HJ. GRADE guidelines: 4. Rating the quality of evidence—Study limitations (risk of bias). *J Clin Epidemiol* 2011, 64:407–715.

12

Adjusting for and Detection of Selection Bias in Randomized Controlled Clinical Trials

Lieven N. Kennes

CONTENTS

12.1 Model... 172
12.2 Performance of the Likelihood Ratio Test—A Simulation Study 180
12.3 Summary... 183
References.. 184

In clinical research, randomization provides the basis for quantitative evaluation of clinical trial data (Rosenberger and Lachin 2002). Because of the lack of representative sampling techniques, randomizing included subjects is indispensable to obtain valid statistical results. Furthermore, randomization is one important technique to avoid bias in a clinical trial. However, only in combination with blinding does randomization actually help to avoid bias (ICH E9 1998). The scientific community as well as regulatory agencies strongly recommend double-blinded clinical trials, where subjects and investigators are unaware of the treatment assignments. Keeping not only the subject but also the investigator blinded throughout the whole trial (masking) is an important feature of a clinical trial, which is extensively discussed in previous chapters of this book. In practice, many clinical trials are not double-blinded. Some situations do not allow for blinding the investigator, while especially late phase clinical trials (phase IIIb/IV) often abstain from blinding to reflect clinical routine.

In the absence of masking, even in randomized clinical trials, third-order selection bias (Berger 2005b, Chapter 5) can substantially influence the study results. Depending on the randomization procedure, third-order selection bias might influence the test decision of a clinical trial (Berger 2005a; Kennes et al. 2011; Proschan 1994). Advanced randomization procedures minimize the impact of third-order selection bias (Berger et al. 2003), but despite their availability, the most popular and most widely used randomization procedure in clinical trials remains the permuted block design.

Thus, in practice, many clinical trials not only use a permuted block design for randomization, but also do not mask past treatment assignments. In combination, these clinical trials become very susceptible for third-order selection bias. To prevent invalid trial results, it is possible to test for the existence of third-order selection bias in a clinical trial and, if detected, adjust for it.

Berger and Exner (1999) suggest testing for selection bias by examining the effect on the response variable of the probability that a subject receives the active treatment.

In a similar way, Berger (2005c) proposes to detect and adjust for selection bias using the reverse propensity score, that is, the probability, conditional on all previous allocations and the allocation procedure, that a given subject will receive a given treatment. Even under the influence of selection bias, comparability of the covariate distributions are ensured within levels of the reverse propensity score.

Ivanova et al. (2005) describe a likelihood ratio test for the presence of selection bias and a test of the treatment effect, adjusted for selection bias. They assume a Blackwell–Hodges model (Blackwell and Hodges 1957) and a binary outcome. By simulation, they demonstrate that the test is approximately chi-square on one degree of freedom. Kennes et al. (2015) validate the distribution of the Ivanova, Barrier, and Berger test theoretically, assuming normally distributed responses and provide a method for confidence interval estimation.

This chapter will address in detail the method developed by Kennes et al. (2015) to detect and correct for selection bias in clinical trials. The mathematical model including the biasing policy is described in Section 12.1. The performance of the introduced test to detect and adjust for selection bias is investigated in a simulation study in Section 12.2.

12.1 Model

In our model, n subjects enroll in a balanced two-arm unmasked clinical trial comparing a control treatment (C) with a new or active treatment (T). After enrollment, subjects are randomly allocated in one of the two treatment groups. Assuming permuted block randomization, we can describe the randomization procedure as a constrained symmetric random walk $\left(R'_t\right)_t$. Let M denote the number of blocks and $2k_j$ the length of block j, $j \in \{1,...,M\}$. Let $\xi_1,..., \xi_n$ be the randomization sequence, where $\xi_i = 1$ if the ith subject is assigned to treatment (T) and -1 if (C). The unconstrained symmetric random walk $(R_t)_t$, with $R_t = \sum_{i=1}^{t} \xi_i$, is constrained on

$$R_t = 0, \text{ for all } t = \sum_{l=1}^{j} 2k_l, \; j = 1,\dots, M.$$

For $M = 1$, we have the special case of Random Allocation Rule.

Because of the lack of masking past assignments, at any time in the trial, the investigator knows the course of the random walk. After i_0 assignments, he knows the number of assignments to group (T), denoted by $N_T(i_0)$, and (C), denoted by $N_C(i_0)$. Without knowing the block size, because of balance at the end of each block, the investigator knows which treatment is more likely to be assigned next, depending on the ratio $N_T(i_0)/N_C(i_0)$. Consciously or subconsciously choosing subjects with certain characteristics for the next assignment is called *convergent strategy* (Blackwell and Hodges 1957) and leads to the following mean structure: Let $Y_1,\dots,Y_n$ be the responses of subjects to treatment, which we assume are normally distributed with variance σ^2 and mean structure

$$E\left(Y_{i+1}\big|\xi_1,\dots,\xi_i\right)=\begin{cases} \mu_g - \eta, & \text{if } P\left(\xi_{i+1}=1\big|\xi_1,\dots,\xi_i\right)>0.5 \Leftrightarrow N_T(i_0)<N_C(i_0), \\ \mu_g, & \text{if } P\left(\xi_{i+1}=1\big|\xi_1,\dots,\xi_i\right)=0.5 \Leftrightarrow N_T(i_0)=N_C(i_0), \\ \mu_g + \eta, & \text{if } P\left(\xi_{i+1}=1\big|\xi_1,\dots,\xi_i\right)<0.5 \Leftrightarrow N_T(i_0)>N_C(i_0). \end{cases}$$

$i = 1,\dots, n - 1, g = 1,2$, for the control and experimental groups, respectively. For the very first assignment in a trial, naturally $E(Y_1) = \mu_g$, because at the beginning of the trial, we have $N_T(i_0) = N_C(i_0) = 0$.

Under this model, η represents the bias accrued by selecting a subject more/less likely to respond to treatment, when the active/control treatment has a higher probability to be assigned next. The appropriate interpretation of the magnitude of η is with respect to σ. This quotient η/σ is called *selection effect*. The magnitude of the selection effect η/σ has not been studied in clinical trials, to our knowledge. In the literature, for η/σ, Proschan (1994) uses the interval $[0.1,\dots,0.5]$, Berger et al. (2003) use the interval $[0, 0.5,\dots,2]$, and Follmann and Proschan (1994) assume a value of 1.

In our model, we assume η to be positive and lower values of the response variable Y are regarded as better. These parameter settings illustrate a scenario where subjects with *better* characteristics are chosen for the next assignment, if the experimental group has a higher probability to be assigned next. Such a behavior might occur, for example, if the investigator wants to test the new procedure on "easier" subjects. One notes that the situation could be reverse and investigators might, again consciously or subconsciously, choose particularly sick or otherwise "difficult" subjects for the experimental group due to the belief in its superior performance. An appropriate modification of the above model is straightforward.

The adjusted test is based on the method of maximum likelihood. With sufficient information, it is possible to estimate the parameter η as well as the unbiased treatment effects μ_1 and μ_2 and the unknown variance σ^2. Sufficient information means knowing not only the treatment assignments, that is, the realizations of $\xi_1, \ldots, \xi_n$, but also their correct order/the randomization protocol, that is, the realization of the random walk $\left(R'_t\right)$. Conducting a clinical trial, this knowledge is available; however, information given in medical publications is generally insufficient.

We define the random variable S_i as

$$S_i = \text{sign}\left(R'_{i-1}\right) = \begin{cases} -1, & \text{if } R'_{i-1} < 0, \\ 0, & \text{if } R'_{i-1} = 0, \\ 1, & \text{if } R'_{i-1} > 0. \end{cases}$$

whereas R'_{i-1} is the value of the constrained random walk after $i-1$ assignments.

The conditional density function of the response Y_i of the ith subject is given by

$$f^{(\theta)}_{Y_i|(\xi_i, S_i)}\left(y_i \mid \tilde{\xi}_i, s_i\right) = \frac{1}{\sqrt{2\pi\sigma^2}} \exp\left(-\frac{1}{2\sigma^2}\left[y_i - \left(\frac{1-\tilde{\xi}_i}{2}\mu_1 + \frac{1+\tilde{\xi}_i}{2}\mu_2 + s_i\eta\right)\right]^2\right),$$

with $\theta := (\mu_1, \mu_2, \eta, \sigma)^T$ and $\tilde{\xi}_i, s_i$ the realizations of ξ_i, S_i.

Thus, the conditional expected response of subject i based on group assignment and randomization history is given in Table 12.1.

TABLE 12.1

Expected Responses of Subject i Based on Group Assignment and Randomization History

	$S_i = -1$	$S_i = 0$	$S_i = 1$
$\xi_i = -1$	$\mu_1 - \eta$	μ_1	$\mu_1 + \eta$
$\xi_i = 1$	$\mu_2 - \eta$	μ_2	$\mu_2 + \eta$

Source: Kennes, L. N., Rosenberger, W. F., Hilgers, R. D.: Inference for blocked randomization under a selection bias model. *Biometrics.* 2015. 71. 979–984. Copyright Wiley-VCH Verlag GmbH & Co. KGaA. Reproduced with permission.

In other words, we have

$$E\left(Y_i | \xi_i, S_i\right) = \frac{1-\xi_i}{2}\mu_1 + \frac{1+\xi_i}{2}\mu_2 + S_i\eta.$$

As the conditional expected response is known but differs among subjects, the $Y_1,...,Y_n$ are independent and not identically distributed (i.n.i.d.) random variables. For a fixed n, the setting can thus be described by a multiple linear regression model

$$Y = Xb + e$$

with

$$Y = \begin{pmatrix} Y_1 \\ \vdots \\ Y_n \end{pmatrix}, \; X = \begin{pmatrix} \frac{1-\tilde{\xi}_1}{2} & \frac{1+\tilde{\xi}_1}{2} & S_1 \\ \vdots & \vdots & \vdots \\ \frac{1-\tilde{\xi}_n}{2} & \frac{1+\tilde{\xi}_n}{2} & S_n \end{pmatrix}, \; b = \begin{pmatrix} \mu_1 \\ \mu_2 \\ \eta \end{pmatrix}, \; \text{and } e \sim N(0, \sigma^2 I_n),$$

I_n denoting the identity matrix of order n.

Given the above linear model, the maximum likelihood estimator (MLE) $\hat{\theta}_n$ for the true parameter $\theta := (\mu_1, \mu_2, \eta, \sigma)^T$ is defined in the usual manner as the value θ that maximizes the likelihood function L_n defined by

$$L_n(\theta) := L\left(\mu_1, \mu_2, \eta, \sigma | y_1, ..., y_n, \tilde{\xi}_1, ..., \tilde{\xi}_n, s_1, ..., s_n\right)$$

$$= \prod_{i=1}^{n} \frac{1}{\sqrt{2\pi\sigma^2}} \exp\left(-\frac{1}{2\sigma^2}\left[y_i - \left(\frac{1-\tilde{\xi}_i}{2}\mu_1 + \frac{1+\tilde{\xi}_i}{2}\mu_2 + s_i\eta\right)\right]^2\right),$$

that is, $L_n(\hat{\theta}_n) \geq L_n(\theta)$, $\forall \theta \in \Theta$, where Θ is the parameter space. Because of the strict monotonicity of the natural logarithm, to find the MLE, we consider the log-likelihood function l_n defined by

$$l_n(\theta) := \ln\left(L_n(\theta)\right) = -\frac{n}{2}\ln(2\pi\sigma^2) - \frac{1}{2\sigma^2}\|Y - Xb\|^2.$$

If the constrained random walk runs only on one side of the x axis and returns to its origin after every second step, $X^T X$ is singular and the MLE corresponding to the likelihood function does not exist, in that $\hat{\theta}_n$ maximizing $L_n(\theta)$ is not unique. In all other cases, it can be shown* that X^TX is not singular and the MLE

$$\hat{\theta}_n = \left(\hat{\mu}_{1,\mathrm{ML}}, \hat{\mu}_{2,\mathrm{ML}}, \hat{\eta}_{\mathrm{ML}}, \hat{\sigma}_{\mathrm{ML}}\right)^T$$

corresponding to the likelihood function by solving the normal equation

$$(X^TX)b = X^TY$$

is given by

$$\hat{\eta}_{\mathrm{ML}} = \frac{a_1a_4 + a_3a_6 - na_2}{a_4^2 + a_6^2 - na_5}$$

$$\hat{\mu}_{1,\mathrm{ML}} = \frac{1}{n}(a_1 - a_3) - \frac{1}{n}(a_4 - a_6)\hat{\eta}_{\mathrm{ML}}$$

$$\hat{\mu}_{2,\mathrm{ML}} = \frac{1}{n}(a_1 + a_3) - \frac{1}{n}(a_4 + a_6)\hat{\eta}_{\mathrm{ML}}$$

$$\hat{\sigma}_{\mathrm{ML}} = \frac{1}{\sqrt{n}}\|Y - Xb\|$$

with

$$a_1 = \sum_{i=1}^n y_i, \; a_2 = \sum_{i=1}^n y_i s_i, \; a_3 = \sum_{i=1}^n y_i \tilde{\xi}_i, \; a_4 = \sum_{i=1}^n s_i, \; a_5 = \sum_{i=1}^n s_i^2, \; a_6 = \sum_{i=1}^n \tilde{\xi}_i s_i.$$

The estimators $\hat{\mu}_{1,\mathrm{ML}}$, $\hat{\mu}_{2,\mathrm{ML}}$ and $\hat{\eta}_{\mathrm{ML}}$ are best linear unbiased estimators (BLUEs) and the estimator $\widehat{\sigma^2}_{\mathrm{ML}} = (\hat{\sigma}_{\mathrm{ML}})^2$ is biased, as expected:

$$E\left(\widehat{\sigma^2}_{\mathrm{ML}}\right) = \frac{n-3}{n}\sigma^2.$$

Even without conducting the below introduced adjusted tests, these results are essentially important on their own. They allow us to not only estimate

---

* A formal proof is given in Kennes (2013), Chapter 3, pp. 36–37.

the real, unbiased treatment effects, but also the selection effect η/σ. Thus, the above estimates yield a first method to check, quantify, describe, and discuss selection bias in a clinical trial.

After quantifying possible effects, we are now interested in the two hypotheses

$$(1) \quad H_0: \mu_1 = \mu_2 \leftrightarrow H_1: \mu_1 \neq \mu_2,$$

$$(2) \quad H_0: \eta = 0 \leftrightarrow H_1: \eta \neq 0.$$

We will use a likelihood ratio test to test the above hypotheses of interest.

Let D be a suitable compact subset of $\mathbb{R}^3$ (see Remark 3.4 in Kennes 2013). Both null hypotheses above can be expressed by embeddings generating a hyperplane in Θ:

$$\kappa_1: D \to \Theta: (x,y,z)^T \to (x,x,y,z)^T,$$

$$\kappa_2: D \to \Theta: (x,y,z)^T \to (x,y,0,z)^T,$$

Regarding embedding κ_i and with

$$\Lambda_{i,n} = \frac{\sup_{\theta \in \Theta} L_n(\theta)}{\sup_{\theta \in \kappa_i(D)} L_n(\theta)}$$

the corresponding likelihood ratio test statistic $2\ln(\Lambda_{i,n})$ yields a quadratic form of random variables. Under certain regularity conditions that are demonstrated below, these random variables are asymptotic normal, which yields

$$2\ln(\Lambda_{i,n}) \overset{d}{\longrightarrow} X, \ X \sim \chi^2_{4-3} = \chi^2_1.$$

Likelihood ratio tests checking whether $\mu_1 = \eta$, $\sigma = \sigma_0$, or $\mu_1 = \mu_2 + \sigma$ etc., could directly be constructed analogously but are not of interest. Hypothesis (2) is a test of selection bias and thus a method to detect selection bias at a certain significance level. Hypothesis (1) is a test of the true treatment difference, adjusting for selection bias.

We now want to discuss the regularity conditions. The following two theorems yield the asymptotic chi-squared distribution on one degree of freedom for the distribution of the above test statistic $2\ln(\Lambda_{i,n})$. For the proofs of both theorems, we refer to the web appendix of Kennes et al. (2015).

Theorem 1

Let Θ be a compact set. In the nondegenerate case, the sequence $(\hat{\theta}_n)_n$, $\hat{\theta}_n = (\hat{\mu}_{1,\mathrm{ML}}, \hat{\mu}_{2,\mathrm{ML}}, \hat{\eta}_{\mathrm{ML}}, \hat{\sigma}_{\mathrm{ML}})^T$, is consistent, that is

$$\hat{\theta}_n \xrightarrow{P} \theta_0.$$

Theorem 2

Let Θ be a compact set. In the nondegenerate case, for $\hat{\theta}_n$ of Theorem 1, we have

$$\sqrt{n}(\hat{\theta}_n - \theta_0) \xrightarrow{d} N\left(0, \bar{\Gamma}^{-1}(\theta_0)\right),$$

where $\bar{\Gamma}(\theta_0)$ is the asymptotic variance–covariance matrix.

The asymptotic variance–covariance matrix is given by

$$\bar{\Gamma}_n(\theta) \to \bar{\Gamma}(\theta),$$

with

$$\Gamma_{ni}(\theta) := -E\left[\ddot{\Phi}_{ni}(Y_{ni}, \theta)|\theta\right],$$

$$\Phi_{ni}(y, \theta) := \mathrm{Hess}\ \Phi_{ni}(y, \theta) = \left(\frac{\partial^2}{\partial\theta_a\partial\theta_b}\Phi_{ni}(y, \theta)\right)_{(a,b)\in\{1,\ldots,4\}\times\{1,\ldots,4\}},$$

and

$$\Phi_{ni}(y, \theta) := \ln\left(f_{ni}^{(\theta)}(y)\right),$$

for a triangular array $\{Y_{ni}, 1 \le i \le n, n \ge 1\}$.
    The likelihood ratio test can be written explicitly as

$$2\ln(\Lambda_{i,n}) = 2n\ln\left(\frac{\hat{\sigma}_{\mathrm{ML}}^{(0)}}{\hat{\sigma}_{\mathrm{ML}}}\right),$$

where $\hat{\sigma}_{\mathrm{ML}}$ is the above MLE for the standard deviation. $\hat{\sigma}_{\mathrm{ML}}^{(0)}$ is computed under the regarded null hypothesis. For Hypothesis (1), regarding the regression model under embedding κ_1 and considering the solution of the respective normal equations, we have

$$\hat{\sigma}_{ML}^{(0)} = \frac{1}{\sqrt{n}} \| Y - X_1 \hat{b}_1 \|,$$

with

$$\hat{b}_1 = \left(\frac{1}{n} a_1 - \frac{1}{n} a_4 \left(\frac{na_2 - a_1 a_4}{na_5 - a_4^2} \right), \frac{na_2 - a_1 a_4}{na_5 - a_4^2} \right)^T \text{ and } X_1 = \begin{pmatrix} 1 & s_1 \\ \vdots & \vdots \\ 1 & s_n \end{pmatrix}.$$

For Hypothesis (2), regarding the regression model under embedding κ_2 and considering the solution of the respective normal equations, we have

$$\hat{\sigma}_{ML}^{(0)} = \frac{1}{\sqrt{n}} \| Y - X_2 \hat{b}_2 \|$$

with

$$\hat{b}_2 = \left(\frac{1}{n}(a_1 - a_3), \frac{1}{n}(a_1 + a_3) \right)^T \text{ and } X_2 = \begin{pmatrix} \dfrac{1 - \tilde{\xi}_1}{2} & \dfrac{1 + \tilde{\xi}_1}{2} \\ \vdots & \vdots \\ \dfrac{1 - \tilde{\xi}_n}{2} & \dfrac{1 + \tilde{\xi}_n}{2} \end{pmatrix}.$$

The acceptance region of the likelihood ratio test with κ_1 yields an asymptotic $(1 - \alpha)$ confidence interval for true treatment difference $\mu_2 - \mu_1$:

$$\left[\frac{\sum_{i=1}^{n} \beta_{1,i} \beta_{2,i}}{\sum_{i=1}^{n} \beta_{2,i}^2} - \sqrt{l}, \ \frac{\sum_{i=1}^{n} \beta_{1,i} \beta_{2,i}}{\sum_{i=1}^{n} \beta_{2,i}^2} + \sqrt{l} \right] \tag{12.1}$$

with

$$\beta_{1,i} = y_i - \frac{a_1 a_5 - a_2 a_4}{na_5 - a_4^2} - s_i \frac{na_2 - a_1 a_4}{na_5 - a_4^2},$$

$$\beta_{2,i} = \frac{1}{2}\left(\frac{a_4 a_6}{na_5 - a_4^2} - 1\right) + \frac{1+\tilde{\xi}_i}{2} - \frac{s_i}{2}\frac{na_6}{na_5 - a_4^2},$$

$$l = \frac{n\hat{\sigma}_{ML}^2 \exp\left(\chi_{1,1-\alpha}^2/n\right)}{\sum \beta_{2,i}^2} + \left(\frac{\sum \beta_{1,i}\beta_{2,i}}{\sum \beta_{2,i}^2}\right)^2 - \frac{\sum \beta_{1,i}^2}{\sum \beta_{2,i}^2}.$$

Analogously, the acceptance region of the likelihood ratio test with κ_2 yields an asymptotic $(1 - \alpha)$ confidence interval for η, describing the magnitude of selection bias:

$$\left[\frac{\sum_{i=1}^{n}\beta_{3,i}\beta_{4,i}}{\sum_{i=1}^{n}\beta_{4,i}^2} - \sqrt{\tilde{l}}, \ \frac{\sum_{i=1}^{n}\beta_{3,i}\beta_{4,i}}{\sum_{i=1}^{n}\beta_{4,i}^2} + \sqrt{\tilde{l}}\right], \tag{12.2}$$

with

$$\beta_{3,i} = y_i - \frac{1}{n}(a_1 + \tilde{\xi}_i a_3),$$

$$\beta_{4,i} = s_i - \frac{1}{n}a_4 - \frac{1}{n}\tilde{\xi}_i a_6$$

$$\tilde{l} = \frac{n\hat{\sigma}_{ML}^2 \exp\left(\chi_{1,1-\alpha}^2/n\right)}{\sum \beta_{4,i}^2} + \left(\frac{\sum \beta_{3,i}\beta_{4,i}}{\sum \beta_{4,i}^2}\right)^2 - \frac{\sum \beta_{3,i}^2}{\sum \beta_{4,i}^2}.$$

12.2 Performance of the Likelihood Ratio Test—A Simulation Study

We can use the above results to estimate the true treatment effects μ_1, μ_2 as well as the selection bias parameter η. Furthermore, we can use the derived likelihood ratio test to not only test for and thus detect selection bias

[Hypothesis (2)], but also conduct an adjusted test for true treatment difference [Hypothesis (1)]:

$$(1) \quad H_0: \mu_1 = \mu_2 \leftrightarrow H_1: \mu_1 \neq \mu_2,$$

$$(2) \quad H_0: \eta = 0 \leftrightarrow H_1: \eta \neq 0.$$

We now want to investigate the performance of the likelihood ratio tests
for both hypotheses. For comparative purposes, $\sigma = 1$ is used in all simulations. Three different sample sizes are chosen: A small trial is simulated
with six blocks of length 10, yielding a sample size of $n = 60$. A second trial
consisting of $n = 150$ subjects with blocks (10, 20, 30, 40, 50) and a larger trial,
including $n = 300$ subjects, is composed of three blocks of length 40, followed
by three blocks of length 60.

The number of replications is 1,000,000. Results for different parameter settings μ_1, μ_2 and η are shown in Table 12.2.

TABLE 12.2

Simulated Type I Error Rate and Power of Likelihood Ratio Tests under Hypotheses
(1) and (2), with Comparison to the t-Test

	Scenario	n	Hypothesis (1)	t-Test	Hypothesis (2)
(T1)	$\mu_1 = 0.5$	60	0.0571	0.0727	0.2202
	$\mu_2 = 0.5$	150	0.0530	0.0725	0.5340
	$\eta = 0.2$	300	0.0513	0.0813	0.8519
(T2)	$\mu_1 = 0.5$	60	0.572	0.1827	0.8114
	$\mu_2 = 0.5$	150	0.0524	0.1754	0.9973
	$\eta = 0.5$	300	0.0512	0.2252	>0.9999
(T3)	$\mu_1 = 0.7$	60	0.1193	0.1671	0.0971
	$\mu_2 = 0.5$	150	0.2267	0.3030	0.1801
	$\eta = 0.1$	300	0.4012	0.5123	0.3286
(T4)	$\mu_1 = 0.5$	60	0.1194	0.0819	0.0976
	$\mu_2 = 0.7$	150	0.2277	0.1657	0.1791
	$\eta = 0.1$	300	0.4013	0.3050	0.3283
(T5)	$\mu_1 = 0.7$	60	0.1193	0.1186	0.0572
	$\mu_2 = 0.5$	150	0.2276	0.2302	0.0525
	$\eta = 0$	300	0.4017	0.4083	0.0516
(T6)	$\mu_1 = 0.5$	60	0.1192	0.1186	0.0572
	$\mu_2 = 0.7$	150	0.2265	0.2296	0.0527
	$\eta = 0$	300	0.4018	0.4081	0.0513

Source: Kennes, L. N., Rosenberger, W. F., Hilgers, R. D.: Inference for blocked randomization
under a selection bias model. *Biometrics*. 2015. 71. 979–984. Copyright Wiley-VCH
Verlag GmbH & Co. KGaA. Reproduced with permission.

Note: Simulations based on 1,000,000 replications with $\sigma = 1$.

In settings (T1) and (T2), the null hypothesis H_0 ($\mu_1 = \mu_2$) holds; however, there is the presence of selection bias ($\eta > 0$). The fourth column demonstrates the performance of the usual two-sided t-test with pooled variance. Because of the presence of selection bias, the type I error rate, here estimated by the relative frequency of decisions for H_1, is inflated. A systematic error occurred, which is larger in (T2) than in (T1) because of a larger selection effect η/σ [(T2): $\eta = 0.5$, (T1): $\eta = 0.2$]. As described in Kennes et al. (2011), we see that an increase in the number of blocks M elevates type I error rate stronger than an increase in the sample size n. For $n = 60$ and $M = 6$, the type I error rate is higher than for $n = 150$, $M = 5$. The third column demonstrates the performance of the above likelihood ratio test with Hypothesis (1). Even when the selection effect is as large as in (T2), the LR-test corrects for selection bias and is able to test equality of the true, unbiased treatment effects. For the likelihood ratio test, we see that as n becomes larger, the type I error rate becomes closer to 0.05. Systematic type I error rate inflation does not occur; that is, even under a strong influence of selection bias, the LR-test asymptotically holds the nominal significance level.

In (T3) and (T5), the active treatment is actually superior ($\mu_1 > \mu_2$), while in (T4) and (T6), the control treatment is superior ($\mu_1 < \mu_2$). Conducting two-sided tests on treatment difference, in all four situations, we now consider power instead of type of error rate. Because of our model specifications, group (T) will always benefit from selection bias while it is the reverse in group (C). The fourth column, where the t-test is conducted, demonstrates this mechanism: Although the absolute difference $|\mu 1 - \mu 2| = 0.2$ was chosen to be the same in (T3) and (T4), in (T4), when the control treatment is superior, power is substantially lower than in (T3) when the active treatment is superior. When (C) is superior, selection bias will work in favor of (T), making observed average responses more equal, while when (T) is superior, selection bias, again working in favor of (T), makes the observed average difference even larger. This asymmetric behavior is not demonstrated by the LR-test in column 3: Also dealing with power, it corrects for selection bias and has equal power for both settings (T3) and (T4). The two-sided LR-test is symmetric as it should be. Furthermore, clearly larger n yields overall higher power.

Scenarios (T5) and (T6) again deal with power, but in absence of selection bias ($\eta = 0$). Here, we see a disadvantage of the LR-test compared to the two-sided t-test. In absence of selection bias, conducting the LR-test yields no gain correcting for selection bias, but is accompanied by a slight loss of power for $n = 150$ and $n = 300$.

The second LR-test introduced in Section 12.1 (with embedding κ_2) tests for selection bias. Column 5 of Table 12.2 demonstrates how powerful the LR-test with Hypothesis (2) is to detect selection bias. A high selection effect with $\eta = 0.5$ is detected almost every time for $n = 150$ and $n = 300$. For $n = 60$, power is still above 80%. For $n = 300$, even a selection effect of 0.2 is detected in 85% of the simulation runs, a small selection effect with $\eta = 0.1$ still in 33%.

TABLE 12.3

Coverage Probabilities for Confidence Intervals (CIs) calculated by Equation 12.1, Equation 12.2, and the Standard Formula for the Difference of Two Means, Related to the *t*-Test

$(\mu_1, \mu_2, \eta, \sigma)$	(0.5,0.5,0.2,1)	(0.7,0.5,0.2,0.5)	(0.5,0.7,0,2)
Equation 12.1	0.9484	0.9488	0.9487
Standard CI related to *t*-test	0.9187	0.8326	0.9502
Equation 12.2	0.9488	0.9488	0.9489

While power is still moderate for $n = 150$ and $\eta = 0.2$, all other scenarios with $n \in \{60, 150\}$ yield insufficient power for small selection effects ($\eta \in \{0.1, 0.2\}$).

In (T5) and (T6), the test decides incorrectly for H_1 in approximately 5% of the simulation runs and hence also asymptotically holds the significance level for Hypothesis (2).

We further investigated the coverage probabilities of the confidence intervals given in Equations 12.1 and 12.2 for $n = 300$, $w = 1{,}000{,}000$ and different parameter settings for $(\mu_1, \mu_2, \eta, \sigma)$. Results are shown in Table 12.3.

We see from Table 12.3 that in each case that we simulated, coverage probabilities were within 0.0016 of 0.95. For comparative reasons, the coverage probabilities of the standard confidence interval for the difference of two means related to the *t*-test were determined as well. Under the influence of selection bias, their confidence levels are substantially lower than the ones determined with Equation 12.1, correcting for selection bias.

12.3 Summary

In randomized clinical trials, selection bias is often not regarded as a threat to clinical trial results. It is widely believed that randomization avoids selection bias. Previous chapters of this book as well as the literature (Berger 2005a,b) demonstrate the opposite. Especially in the event of unmasked treatment assignments, third-order selection bias might strongly influence the test decision of a clinical trial (Kennes et al. 2011; Proschan 1994).

In this chapter, for a continuous endpoint, statistical hypotheses tests to detect and correct for selection bias are developed with the method of maximum likelihood. These tests can easily be conducted on clinical trial data and yield direct information on the presence and magnitude of selection bias as well as results on the true treatment difference adjusted for selection bias. Performing these tests and deriving the corresponding confidence intervals will facilitate the ascertainment of selection bias and its impact on the treatment effect, especially in unmasked randomized clinical trials.

References

Berger, V. W. (2005a). Quantifying the magnitude of baseline covariate imbalances resulting from selection bias in randomized clinical trials. *Biometrical Journal* 47, 119–127.

Berger, V. W. (2005b). *Selection Bias and Covariate Imbalances in Randomized Clinical Trials.* Chichester: Wiley.

Berger. V. W. (2005c). The reverse propensity score to detect selection bias and correct for baseline imbalances. *Statistics in Medicine* 24, 2777–2787.

Berger, V. W. and Exner, D. V. (1999). Detecting selection bias in randomized clinical trials. *Controlled Clinical Trials* 20, 319–327.

Berger V. W., Ivanova, A., and Knoll, M. (2003). Minimizing predictability while retaining balance through the use of less restrictive randomization procedures. *Statistics in Medicine* 42, 3017–3028.

Blackwell, D. and Hodges, J. (1957). Design for the control of selection bias. *Annals of Mathematical Statistics* 25, 449–460.

Follmann, D. and Proschan, M. (1994). The effect of estimation and biasing strategies on selection bias in clinical trials with permuted blocks. *Journal of Statistical Planning and Inference* 39, 1–17.

ICH E9. 1998. Statistical principles for clinical trials. Current version dated February 5, 1998. http://www.ich.org. Accessed August 2017.

Ivanova, A., Barrier, R. C., and Berger, V. W. (2005). Adjusting for observable selection bias in block randomized trials. *Statistics in Medicine* 24, 1537–1546.

Kennes, L. N., Cramer, E., Hilgers, R. E., and Heussen, N. (2011). The impact of selection bias on test decisions in randomized clinical trials. *Statistics in Medicine* 30, 2573–2581.

Kennes, L. N. (2013). The Effect of and Adjustment for Selection Bias in Randomized Controlled Clinical Trials. Aachen, Germany: RWTH Aachen University (doctoral dissertation).

Kennes, L. N., Rosenberger, W. F., Hilgers, R. D. (2015). Inference for blocked randomization under a selection bias model. *Biometrics* 71, 979–984.

Proschan, M. (1994). Influence of selection bias on type I error rate under random permuted block designs. *Statistica Sinica* 4, 219–231.

Rosenberger, W. F. and Lachin, J. M. (2002). *Randomization in Clinical Trials: Theory and Practice.* New York: Wiley.

13

Randomization and the Randomization Test: Two Sides of the Same Coin

Patrick Onghena

CONTENTS

13.1 Randomization 186
    13.1.1 Randomization in Clinical Trials.. 186
    13.1.2 Why Randomize? 187
    13.1.3 Clarification of the Equating Property of Randomization..... 188
    13.1.4 An Example.. 189
    13.1.5 Concluding Remarks on the Equating Property
          of Randomization.. 190
13.2 The Randomization Test.. 191
    13.2.1 Notation.. 192
    13.2.2 The Null Hypothesis and Several Possibilities
          for the Alternative Hypothesis.. 193
    13.2.3 The Test Statistic 195
    13.2.4 The Randomization Test p Value... 195
    13.2.5 An Example.. 197
    13.2.6 Main Features of the Randomization Test............................... 199
          13.2.6.1 Validity and Power by Construction......................... 199
          13.2.6.2 Flexibility with Respect to the Design
                   and the Test Statistic 200
          13.2.6.3 No Random Sampling and No Distributional
                   Assumptions Required.. 201
          13.2.6.4 Frequentist, Conditional, and Causal Inference 201
13.3 Two Sides of the Same Coin... 202
References.. 203

The idea of randomization for eliminating bias in empirical research and the proposal for using the corresponding technique of a randomization test are relatively recent methodological developments, taking into account the long history of the scientific enterprise (David 1995; Folks 1984; Hacking 1988; Salsburg 2001; Stigler 1978, 1986). In the nineteenth century, randomization was introduced in psychophysical investigations to keep participants unaware of the experimental manipulations and thus avoiding systematic distortions due to participants' knowledge or their willingness to act (or not) according to the researcher's hypotheses (see, e.g., Peirce & Jastrow 1885), but it was only in the first half of the previous century that Fisher (1925, 1926, 1935) convincingly started promoting randomization as an essential ingredient of any proper experimental research plan and as a way to validate statistical tests.

In medical research, Sir Bradley Hill has been given credit for setting the randomized controlled trial as the gold standard for comparing treatment effects, with the streptomycin study in 1948 as the exemplar (Cochrane 1972; Matthews, this volume). Nowadays, randomization belongs to the canon of valid empirical research in evidence-based medicine (Begg et al. 1996; Higgins & Green 2011; Jadad & Enkin 2007). However, from its inception, and continued throughout the second half of the previous century, (mainly Bayesian) statisticians have seriously questioned the use of randomization, its methodological benefits, and its implications for statistical analysis (Basu 1975, 1980; Kadane & Seidenfeld 1990; Lindley 1982; Savage 1962; Stone 1969), with doubts remaining up to this very day (see Saint-Mont 2015, for an overview).

Because randomization is only recently adopted in the realm of science and because experts in methodology and statistics are still debating its virtues and vices, it is no surprise that there remain several misunderstandings, resistance, and suboptimal surrogates to the idea of randomization in scientific research (Berger & Bears 2003; this volume). In this chapter, we want to contribute to the ongoing debate by providing a framework for conceptualizing randomization and by linking randomization to a statistical test. Our aim is to show how randomization works, how it does not work, and why a discussion of randomization without reference to a randomization test is incomplete.

13.1 Randomization

What is randomization and how does it work? Perhaps more importantly: How does it *not* work?

13.1.1 Randomization in Clinical Trials

Given the topic of this book, the presentation will be restricted to randomization in clinical trials. Thus, the experimental units are patients, and the "randomization" refers to the random assignment of patients to the treatments that are compared in the clinical trial (Jadad & Enkin 2007). If the number of patients in the trial is denoted by N, the number of treatments is denoted by J, and the number of patients assigned to treatment $T_1, T_2,... T_j,... T_J$ is denoted by $n_1, n_2, \ldots n_j, \ldots n_J$, respectively, then the total number of possible distinct assignments of patients to treatments, denoted by K, is given by the multinomial coefficient

$$K = \frac{N!}{\prod_{j=1}^{J} n_j}.$$

(13.1)

If there are only two treatments, $J = 2$, for example, in a double-blind clinical trial of a medical agent versus a placebo control, the total number of possible assignments simplifies to a binomial coefficient:

$$K = \binom{N}{n_1}.$$

(13.2)

Randomly assigning, or what Fisher (1935, p. 51) called "the physical act of randomisation," can be conceptualized as taking a simple random sample of size one out of the population of K possible assignments, with the population of possible assignments generated according to the design specifications (Finch 1986; Kempthorne 1986). Equivalently, if the N available patients are considered as the population, the random assignment can be conceptualized as taking a sample without replacement of size N and respecting the order of the sampled patients, the order of the treatments, and the number of patients assigned to the treatments (Efron & Tibshirani 1993; Good 2005).

13.1.2 Why Randomize?

There are many reasons why researchers include randomization in their empirical studies, and reasons may vary between disciplines (Greenberg 1951; Kempthorne 1977; Rubin 1978; Senn 1994). In double-blind clinical trials, randomization is included to safeguard the double-blindness, to be consistent with the principle of equipoise, and to take into account ethical considerations with respect to informed consent and with respect to a fair allocation of decent and affordable health care services (Begg 2015; Freedman 1987; Matthews, this volume).

From a general statistical perspective, there are two closely related reasons given in the literature: the elimination of bias and the justification of a statistical test (Cochran & Cox 1950; Cox & Hinkley 1974; Kempthorne 1952, 1955; Kempthorne & Folks 1971; Rosenberger & Lachin 2015). In this section, we will focus on the elimination of bias. In Section 13.2, we will demonstrate how randomization may justify a statistical test.

The property of bias elimination can be positively reformulated as the equating property of randomization. Randomization is used in clinical trials to equate the groups that are assigned to the different treatments in a statistical way. As Rubin (2008) put it:

> Another reason why randomized experiments are so appealing (...), is that they achieve, in expectation, "balance" on all pre-treatment-assignment variables (i.e., covariates), both measured and unmeasured. Balance here means that within well-defined subgroups of treatment and control units, the distributions of covariates differ only randomly between the treatment and control units (p. 809).

By using randomization, treatments become comparable because the patients are exchangeable. Because systematic differences between the patients can be ignored, the differences between the treatment outcomes can be attributed to the differences between the treatments and not to any differences between the patients accidentally assigned to the treatments.

And here the confusion begins. Researchers might be tempted to use randomization as a magical device that automatically washes out any difference between patient groups in a single clinical trial (see Saint-Mont 2015, for some examples). However, the equating property only holds in a statistical way. In the Rubin (2008) quote, it is important to notice the phrase "in expectation" (and to interpret this expectation in a strict statistical way) and that "the distributions of covariates" still differ "between the treatment and control units," be it "only randomly."

13.1.3 Clarification of the Equating Property of Randomization

What do this "statistical way" and "expectation" mean? They mean that, *on average*, the differences between the treatment averages of any potentially confounding variable are zero. Suppose that we use a completely randomized design with the number of possible assignments of N patients to J treatments equal to K, as given in Equation 13.1. Take any confounding variable Z, with z_{ij} as the measurement of that variable of patient i in treatment j. Randomly assigning patients to treatments implies randomly assigning the measurements of the confounding variable to the treatments, so let z_{ijk} be the measurement of patient i in treatment j for assignment k.

Consider a particular assignment of the N measurements to the J treatments. Then consider all $K - 1$ rearrangements of these measurements to the treatments and organize them in a large $N \times K$ matrix. In such a matrix, the total sum of all measurements is equal to K times the sum of the observed measurements (or any other rearrangement of the measurements) because the sum is a constant for each assignment:

$$\sum_{k=1}^{K}\sum_{j=1}^{J}\sum_{i=1}^{n_j} z_{ijk} = K\sum_{j=1}^{J}\sum_{i=1}^{n_j} z_{ij}. \tag{13.3}$$

The total sum of all measurements in the matrix is also equal to $\dfrac{N}{n_j}$ times the sum of all measurements within a particular treatment in the matrix, z_{ik}, computed over all possible assignments, because all measurements occur an equal number of times within a treatment. The sum of all measurements is partitioned in terms that are proportional to the number of measurements in each treatment:

$$\sum_{k=1}^{K}\sum_{j=1}^{J}\sum_{i=1}^{n_j} z_{ijk} = \frac{N}{n_j}\sum_{k=1}^{K}\sum_{i=1}^{n_j} z_{ik}. \tag{13.4}$$

Working back from Equation 13.4 and substituting for Equation 13.3 gives

$$\sum_{k=1}^{K}\sum_{i=1}^{n_j} z_{ik} = \frac{n_j}{N}\sum_{k=1}^{K}\sum_{j=1}^{J}\sum_{i=1}^{n_j} z_{ijk} = \frac{n_j K}{N}\sum_{j=1}^{J}\sum_{i=1}^{n_j} z_{ij}. \tag{13.5}$$

Consequently, if you randomly select one assignment out of the total number of possible assignments, then the expected value of the average of the measurements within each treatment is equal to the average of the N measurements:

$$E\left(\frac{\sum_{i=1}^{n_j} z_i}{n_j}\right) = \frac{\sum_{k=1}^{K}\sum_{i=1}^{n_j} z_{ik}}{Kn_j} = \frac{\sum_{j=1}^{J}\sum_{i=1}^{n_j} z_{ij}}{N}. \tag{13.6}$$

This implies that, on average, all treatment averages are identical, and in this sense, the groups in a completely randomized design are equated. In other words, the bias due to potentially confounding variables is, "in expectation," eliminated.

13.1.4 An Example

A numerical example may clarify this obvious, but nevertheless easy to misunderstand, property. Suppose that you are testing a new diet against a control and that four patients are available. You use a completely randomized design with $N = 4$, $J = 2$, $n_1 = 2$, $n_2 = 2$, and therefore $K = 6$. The weight of the patients before the trial is evidently a confounding variable; suppose that you try to control this confounding variable by randomization. Commonly, the numerical values of the confounding variables are unknown, but suppose in this case, for the purpose of demonstration, we have put the patients

TABLE 13.1

Hypothetical Example of the Equating Property of Randomization in a Completely
Randomized Design with Two Treatments and Two Patients in Each Treatment

	Treatment		Control		Average
Assignment 1	62	75	76	127	340/4 = 85
Assignment 2	62	76	75	127	340/4 = 85
Assignment 3	62	127	75	76	340/4 = 85
Assignment 4	75	76	62	127	340/4 = 85
Assignment 5	75	127	62	76	340/4 = 85
Assignment 6	76	127	62	75	340/4 = 85
Average	1020/12 = 85		1020/12 = 85		2040/24 = 85

TABLE 13.2

Hypothetical Example of Table 13.1 with Averages for Each
Treatment and the Deviation for Each Assignment

	Treatment	Control	Deviation
Assignment 1	68.5	101.5	−33
Assignment 2	69	101	−32
Assignment 3	94.5	75.5	19
Assignment 4	75.5	94.5	−19
Assignment 5	101	69	32
Assignment 6	101.5	68.5	33
Average	510/6 = 85	510/6 = 85	0/6 = 0

on a scale before the trial is conducted and registered the weights for the four
patients to be 62, 75, 76, and 127 kg, respectively. The matrix of all possible
assignments and the marginal averages are given in Table 13.1. Table 13.2
shows the averages for each treatment and the differences between the averages for each assignment.

It turns out that the average (i.e., the expected value) of the averages of
the measurements for treatment as well as for control is equal to 85. This
is exactly the average of the four numbers 62, 75, 76, and 127 (and also the
overall average of all numbers in the matrix). Notice also in Table 13.2 that
any specific assignment violates the assumed magical equating property:
The deviation between the treatment average and the control average ranges
from 33 to −33 kg. It is only on average that the deviation is zero. Hence, it
is only on average that the equating property of randomization works. It is not
guaranteed to work for any single assignment.

13.1.5 Concluding Remarks on the Equating Property of Randomization

It is sometimes argued that the magical equating property for single assignments *will* work if a larger number of patients are recruited. However, this is

a red herring. Although the deviations for one particular assignment and one particular confounding variable might decrease by increasing the number of patients, the sheer number of potential known and unknown confounding variables and their interactions is close to infinity. Thus, the probability that there will show up an imbalance for at least one confounding variable is close to one.

This clarification of the equating property of randomization and the small numerical example may seem trivial and nonrealistic but at the same time they show the deeper and limited benefits of randomization. Too many handbooks in statistics give only cursory attention to the "physical act" and in this way turn randomization into a design ritual or, even worse, into a reporting ritual that has no bearing on reality. By contrast, our elaboration and Equation 13.6 show that the equating property holds for the very broad class of completely randomized designs, so even for designs that are, what is called, "unbalanced" (i.e., designs with an unequal number of patients in the treatments). Hence, even unbalanced designs "achieve, in expectation, 'balance' on all pre-treatment-assignment variables (i.e., covariates), both measured and unmeasured." In addition, the example illustrates that the property does not need nicely behaving Gaussian or symmetrically distributed confounding variables or asymptotic derivations. The property already works with only two patients in each of two treatments, even with a blatantly skewed confounding variable.

Thus, as a conclusion, the equating property is one of the main reasons that randomization deserves a place in the clinical trials design toolbox. However, we must not get carried away by statistical enthusiasm: This equating property does not guarantee optimal balance and comparability in any particular assignment. Furthermore, if a researcher has prior knowledge about the patients or about the effectiveness of the treatments, then other ways of equating may be used, such as matching or blocking (Campbell & Stanley 1963; Cook & Campbell 1979; Shadish et al. 2002) or using adaptive allocation rules (Berry et al. 2010; Hu & Rosenberger 2006). Just to be sure, randomization was never proposed as a panacea solution for all evils. Randomization is at its best in conjunction with other methodological maneuvers to gain control and complemented by a statistical test based on the randomization as it was actually carried out.

13.2 The Randomization Test

In these discussions it seems to have escaped recognition that the physical act of randomisation, which, as has been shown, is necessary for the validity of any test of significance, affords the means, in respect of any particular body of data, of examining the wider hypothesis in which no normality of distribution is implied (Fisher 1935, p. 51).

In this quote from the groundbreaking *Design of Experiments*, Fisher (1935) is opposing the critics of Student's *t* test who emphasize that the assumption of normally distributed populations represents a serious limitation to its applicability. He continues by proposing a statistical test for data on the difference between the relative growth rates of cross- and self-fertilized corn, collected by Charles Darwin and analyzed by Francis Galton, in which no normally distributed populations are assumed. He calls this a test of "the wider hypothesis" that the two distributions are identical. This test is a precursor of the significance test "which may be applied to samples from any populations" proposed by Pitman (1937a,b, 1938) and which has later been termed a "randomization test" (Berry et al. 2014; David 2008, Kempthorne 1952, 1955), but for our purposes, it suffices to join Fisher (1935) in emphasizing that

- Randomization is needed for the validity of any statistical test.
- The act of randomization is physical.
- There is no need for distributional assumptions if you want a test of the wider hypothesis that the two distributions are identical.

Hence, an account of the purposes of randomization is incomplete without reference to the statistical test for which it was originally devised. Because the equating property of randomization only describes a "statistical" regularity, it needs to be complemented by a statistical test to minimize erroneous scientific inferences based on the observed data; the statistical test minimizes these erroneous scientific inferences to a known (and small) degree. Furthermore, the foundation of the statistical test is an operation of randomization that is actually carried out, not just a mere thought experiment or assumption. Incidentally, this "randomization" test also relieves the researcher from assuming normally distributed populations; it is the distribution-free statistical test *par excellence*.

In this section, we briefly present the rationale of this test based on randomization and discuss its main features. For more extensive presentations and discussion, the reader is referred to Good (2005), Manly (2007), and Edgington and Onghena (2007).

13.2.1 Notation

We will develop the notation for a completely randomized design with a simple treatment-control comparison involving N available patients. However, the notation can easily be extended for completely randomized designs comparing two treatments, for completely randomized designs comparing three treatments or more, or for other designs.

Let $x' = (x_1, \dots x_2, \dots, x_i, \dots, x_N)$ be a design vector with $x_i = 1$ if patient i is assigned to treatment and $x_i = 0$ if patient i is assigned to control. The randomness of the assignment involves the random selection of one particular design vector $x'_k = (x_{1k},\dots, x_{2k},\dots, x_{ik},\dots, x_{Nk})$ from the set of all possible design vectors $X = \{x_1, x_2,\dots, x_k,\dots, x_K\}$. The randomly selected design vector is called the observed design vector and denoted by x_{obs}, and the set of all possible design vectors X is called the reference set. The cardinality of the reference set, K, with n_1 patients assigned to treatment, leaving $N - n_1$ patients for control, was given in Equation 13.2.

If the trial contains a comparison of $J > 2$ treatments, then $J - 1$ design vectors and additional subscripts are needed. It is also possible that the x_i values represent the "dose" of the treatment if the treatment is manipulated quantitatively. In that case, only one design vector is needed, even with $J > 2$. The cardinality of the reference set for a completely randomized design with J treatments was given in Equation 13.1. The cardinality of the reference set for a randomized design in which the treatment is manipulated quantitatively is equal to $N!$ (see Chapters 8 and 9 of Edgington & Onghena 2007 for some interesting applications for designs with quantitatively manipulated treatments).

Furthermore, consider $u' = (u_1, u_2\dots, u_i,\dots, u_N)$, a vector of basic responses, and $y' = (y_1, y_2\dots, y_i,\dots, y_N)$, a vector of actually observed responses. The basic responses are the responses that the patients would give if the patients are assigned to control (c.q. if the treatment has no effect). These basic responses are transformed to actually observed responses, taking into account treatment assignment and potential treatment effects.

13.2.2 The Null Hypothesis and Several Possibilities for the Alternative Hypothesis

The null hypothesis H_0 of the randomization test in a completely randomized design for a comparison of treatment and control states that there is no treatment effect:

$$H_0: y = u. \tag{13.7}$$

In other words, if the null hypothesis is true, we observe the basic responses. This formula has to be slightly adapted or extended if two or more treatments are compared, to take into account that, in that case, the null hypothesis is formulated in terms of "no differential effects of treatments," but this does not alter the basic reasoning and procedure.

Edgington (1986, p. 531) defined the alternative hypothesis of a randomization test as the hypothesis that at least one patient would have responded differently under one of the other treatments. This is consistent with the Fisherian approach, in which the alternative hypothesis is just the complement of the null hypothesis. In the Neyman–Pearson approach, however, one should try to consider more specific alternative hypotheses.

The simplest model for specific alternative hypotheses, following the Neyman–Pearson approach, is that the treatment has a constant additive effect (denoted by Δ) on the responses. This model is called the "unit–treatment additivity model" and it is the model that is most popular and well studied within nonparametric statistics (see, e.g., Cox & Reid 2000; Hinkelmann & Kempthorne 2005, 2008, 2012; Lehmann 1975):

$$H_1: y = u + \Delta x_k. \tag{13.8}$$

For example, if the treatment adds one measurement unit to each basic response if the patient is assigned to the treatment, the null and alternative hypothesis correspond to

$$H_0: \Delta = 0$$
$$H_1: \Delta = 1. \tag{13.9}$$

The unit–treatment additivity model presupposes that there is no unit–treatment interaction: Δ is a constant.

A model that allows for unit–treatment interaction is the multiplicative model. In the multiplicative model, the treatment effect is proportional to the basic responses:

$$H_1: y = u\,(1 + \Delta x_k). \tag{13.10}$$

For example, if one assumes that the treatment doubles each basic response for the patient that receives the treatment, the null and alternative hypothesis correspond to

$$H_0: \tau = 0$$
$$H_1: \tau = 1. \tag{13.11}$$

An infinite number of other models can be conceived. Models that contain extreme unit–treatment interaction are models in which Δ (or τ) is a vector, with Δ_i (or τ_i) varying from $i = 1$ to N. In such a model, each patient has his or her own additive or multiplicative treatment response. It goes without saying that, in most cases, researchers would be very hard-pressed to produce such detail in the alternative hypothesis, with all the Δ_i specified before the trial is conducted. If it were possible to produce such detail, then one could even question the need for conducting the trial in the first place.

The typical Fisherian solution (echoed by Edgington 1986) is to set the lower bound (i.e., all Δ_i equal to zero, except for one) and to consider the complete set of alternative hypotheses in an unrestricted unit–treatment interaction model. An intermediate solution, which can be used for conditional power analysis, is to assume that the Δ_i or τ_i represent a random sample from a certain well-described population distribution of treatment effects, such as the exponential distribution used in Keller's (2012) simulation study to investigate the power of randomization tests when N is small.

13.2.3 The Test Statistic

Like any other statistical test, the randomization test needs a test statistic. Unlike any other statistical test, however, the researcher is free in his choice of this statistic.

A test statistic is needed because the researcher has collected a batch of data, wants to summarize these data, and wants to know whether or not a treatment effect can be inferred from the data summary, possibly also how large this effect is, given the complete batch. This statistic can be chosen freely by the researcher in accordance to his theory about the possible treatment effect (the alternative hypothesis discussed in Section 13.2.2), but for the validity of the procedure, it is of paramount importance that the choice is made independently from knowing the data. Usually, this independence is assured by making the choice before the data are collected, but if appropriate masking techniques are used, then it is also possible to postpone this choice until a moment before the data are disclosed to the researcher or the data analyst (see Ferron & Foster-Johnson 1998; Ferron & Jones 2006; Ferron & Levin 2014, for some interesting applications of data masking).

Formally, the test statistic s is a function of the design vector and the response vector: $s = s(x, y)$. If, for example, the unit–treatment additivity model of Equation 13.8 is assumed in a completely randomized design with n_1 patients assigned to treatment, leaving $N - n_1$ patients for control, then the additive treatment effect on all patients can be captured by a difference between the average response of the treatment and the average response of the control, $s = \dfrac{1}{n_1} x'y - \dfrac{1}{N - n_1}(1' - x')y.$

13.2.4 The Randomization Test p Value

The final step of a statistical test involves the calculation of the p value, which is the probability to obtain a test statistic as extreme as the observed test statistic or even more extreme, conditional on the truth of the null hypothesis. The only random element in the randomization model under the null hypothesis, as it was presented in the previous paragraphs, is given by the random selection of the observed design vector, x_{obs}, out of the reference set, X. Thus, the only way to calculate probabilities under the null hypothesis concerns this random selection.

Define the test statistic $s_k = s(x_k, y)$ as a function of the design vector and the response vector; the design vector is random and the response vector is fixed. Let $s_{obs} = s(x_{obs}, y)$ be the observed value of the test statistic, also called the observed statistic for short, or put $s_{obs} = s_1$ for convenience. The other values of the test statistic $s_2, s_3, \ldots, s_k, \ldots, s_K$ are called the additional reference statistics, and the frequency distribution on the reference statistics is called the reference distribution.

The probabilities of particular subsets of test statistics can easily be calculated from this reference distribution. More specifically, if large values of the test statistic represent effectiveness, then right-tail p values are obtained as

$$p_{\text{right}} = P(s_k \geq s_{obs}) = \frac{1}{K} \sum_{k=1}^{K} I(s_k \geq s_{obs}), \tag{13.12}$$

with $I(a)$ being the indicator function, returning a 1 if proposition a is true and returning 0 otherwise. Remark that the number of reference statistics that are as large as the observed statistic or larger, call this η, is at least 1 because the observed statistic, $s_{obs} = s_1$, is also a reference statistic:

$$\eta = Kp_{\text{right}}$$
$$= \sum_{k=1}^{K} I(s_k \geq s_{obs}) \tag{13.13}$$
$$= 1 + \sum_{k=2}^{K} I(s_k \geq s_1).$$

This implies that the minimal p value of the randomization test is equal to $1/K$.

If small values represent effectiveness, then left-tail p values are obtained:

$$p_{\text{left}} = P(s_k \leq s_{obs}) = \frac{1}{K} \sum_{k=1}^{K} I(s_k \leq s_{obs}). \tag{13.14}$$

Two-tailed p values are obtained by defining a test statistic with an absolute value or with a square and applying Equation 13.12.

Fisher (1925, 1935) used the p value as a measure of evidence against the null hypothesis: the smaller this p value, the more evidence against the null hypothesis. Notice, however, that this p value does not quantify the probability that the null hypothesis is true. It is merely a calculation of the probability to observe the data (or even more extreme data), given that the null hypothesis is true (Edgington 1970; Wasserstein & Lazar 2016).

In the Neyman–Pearson approach, this p value is used in a decision rule to reject or accept the null hypothesis. The null hypothesis is rejected if and only if the p value is not larger than the level of significance α:

$$\text{Reject } H_0 \iff p \leq \alpha. \tag{13.15}$$

Following this approach, the level of significance α is set before the trial is run and is the yardstick that a researcher wants to use as the criterion for (the largest probability of) an improbable event. If a p value is obtained that is equal to this criterion or even smaller, then an improbable event has happened if the null hypothesis is true. Because a researcher does not want to believe in improbable events, he or she rejects the null hypothesis. Traditionally, values for α of 5%, 1%, or 0.1% are used.

13.2.5 An Example

Reconsider the hypothetical example of testing a new diet against a control and now take the patient's weight registered at the end of the trial as the outcome variable. Suppose that a researcher wants to test the null hypothesis that the treatment has no effect, with 10 consecutive patients eligible for the trial, and that he or she chooses the difference between the averages as the test statistic. The level of significance α is set at 5%.

Five patients are randomly assigned to treatment and five patients are randomly assigned to control. Applying Equation 13.2 gives $K = \binom{10}{5} = 252$ design vectors in the reference set, X. These design vectors are

$$x'_1 = (1, 1, 1, 1, 1, 0, 0, 0, 0, 0)$$
$$x'_2 = (1, 1, 1, 1, 0, 1, 0, 0, 0, 0)$$
$$x'_3 = (1, 1, 1, 1, 0, 0, 1, 0, 0, 0)$$

and all other combinations up to

$$x'_{252} = (0, 0, 0, 0, 0, 1, 1, 1, 1, 1).$$

Suppose we randomly selected the first design vector x_1 and that the following response vector is obtained: $y' = (51, 62, 75, 76, 77, 74, 89, 90, 92, 127)$

in kilograms. The observed statistic is $s_{obs} = s_1(x_1, y) = \dfrac{1}{5}\, x'y - \dfrac{1}{5}\, (1' - x')y =$ 68.2 − 94.4 = −26.2. All additional reference statistics are calculated and the only design vectors that would have led to a smaller reference statistic are (1, 1, 1, 1, 0, 1, 0, 0, 0, 0) with a reference statistic of −27.4, (1, 1, 1, 0, 1, 1, 0, 0, 0, 0) with a reference statistic of −27, and (1, 1, 0, 1, 1, 1, 0, 0, 0, 0) with a reference statistic of −26.6. Because there are only three additional reference statistics smaller than or equal to the observed statistic, applying Equation 13.14 gives a p value equal to 4/252 = 0.0159. In clinical trials, it usually cannot be ruled out that the treatment has a paradoxical effect, and therefore, two-tailed tests are preferred. If the absolute value of the difference between the averages is taken as a test statistic, then $s_{obs} = s_1\,(x_1, y) = \left| \dfrac{1}{5} x'y - \dfrac{1}{5}(1' - x')y \right| =$

$|68.2 - 94.4| = 26.2$. Applying Equation 13.12 gives a p value of 8/252 = 0.0317. Hence, the null hypothesis that there is no treatment effect can be rejected at the 5% significance level, because 0.0317 is not larger than 0.05 (Equation 13.15). There appears to be a statistically significant smaller average weight in the treatment as compared to the control. Figure 13.1 shows the reference distribution and indicates the area with the reference statistics that are larger than or equal to the observed statistic, in absolute value.

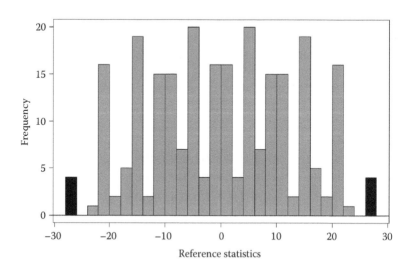

FIGURE 13.1
Reference distribution for a randomization test in a completely randomized design with two treatments and five patients in each treatment. The data are 51, 62, 75, 76, and 77 for one treatment and 74, 89, 90, 92, and 127 for the other, and the difference between the treatment averages is taken as the test statistic. The black bars correspond to the reference statistics that are larger than or equal to the observed statistic, in absolute value. The two-tailed randomization test p value corresponds to 8/252.

13.2.6 Main Features of the Randomization Test

The randomization test has several striking features, which set it apart from other inferential procedures: it is valid and powerful by construction, it can be used with any design and any test statistic, without random sampling and without assuming specific population distributions, and it results in a frequentist, conditional, and causal inference.

13.2.6.1 Validity and Power by Construction

The randomization test is exactly valid by construction. If the null hypothesis is true and virtual repetitions of the trial are contemplated, the response vector is invariant. Thus, whatever design vector would be applied, the responses will always be identical if the null hypothesis is true. Because each design vector corresponds to one reference statistic, randomly selecting a design vector with probability $1/K$ implies randomly selecting a reference statistic with probability $1/K$. Consequently, the probability of a particular subset of reference statistics can be calculated as the proportion of these reference statistics in the finite reference distribution, as is done in Equations 13.12 and 13.14.

If the decision rule of Equation 13.15 is used, then two kinds of errors can be made. An error of the first kind is made when the rule leads to a rejection of a true null hypothesis (a false alarm). An error of the second kind is made when the rule does not lead to a rejection of a false null hypothesis (a missed detection).

It can be shown that the randomization test controls the probability of an error of the first kind in the strict sense, which means that this probability is never larger than the level of significance α (Edgington & Onghena 2007; Onghena 1994). It is possible that the probability of an error of the first kind is substantially smaller than the level of significance α (i.e., it is a conservative test) because the reference distribution is discrete (with probability steps of $1/K$) and because additional reference statistics may be tied with the observed statistic (Keller 2012; Onghena 1994), but this conservatism can be made negligible by setting the level of significance α equal to a multiple of $1/K$ or by adding auxiliary randomization (Berger 2000, 2004, 2007, 2008; Edgington & Onghena 2007; Keller 2012; Onghena 1994). This error rate control holds for randomization tests in all randomized trials, even for trials that have very small N and even if $n_i = 1$ in some or all of the treatments. This feature distinguishes the randomization test from many other inferential procedures that have only guaranteed validity for large N or that need two or more observations in each treatment (because within-treatment variance has to be estimated).

Just like for any other inferential procedure, however, N is a crucial determinant of the probability of an error of the second kind for a randomization test. For an extreme example, if N is chosen such that $K < 20$ by Equation 13.1, then Equation 13.15 implies that a 5% level randomization test will result in an error of the second kind with absolute certainty (probability of 100%) because in that case, it is impossible to have $p \leq \alpha$ for any true alternative hypothesis. Besides N, the statistical power of a randomization test (i.e., the complement of the probability of an error of the second kind) is a function of the design, the vector of basic responses, the treatment effect, the test statistic, and the level of significance α (Gabriel & Hall 1983; Gabriel & Hsu 1983; Keller 2012; Kempthorne & Doerfler 1969). Because the most sensitive test statistic can be devised for the specific treatment effect that is expected, randomization tests with optimal power can be constructed in a variety of situations. For example, in a completely randomized design with two or more than two treatments, if the alternative hypothesis specifies random sampling from normally distributed populations with different population averages, the randomization test using t or F as a test statistic has an asymptotic relative efficiency of 100% as compared to the classical parametric Student's t or F test (Hoeffding 1952). If a test statistic based on ranked transformed data is used (Wilcoxon-type statistics), then the asymptotic relative efficiency of the randomization test compared to Student's t test is never smaller than 0.864 and is larger for logistic, exponential, and double exponential distributions; for Cauchy distributions, the asymptotic relative efficiency is even infinitely large (Hollander et al. 2014; Lehmann & Romano 2005).

13.2.6.2 *Flexibility with Respect to the Design and the Test Statistic*

A distinctive feature of the randomization test is that it can be used with any customized randomized design. Other statistical tests are usually linked to a particular conventional setting such as the one-sample problem, two independent samples, paired samples, the split-plot design, and so on, but a randomization test can be applied to any design that fits the practical circumstances of the study under consideration, even if these circumstances call for an unconventional randomization procedure. For example, because of practical or ethical considerations, it might be needed to randomly assign only one patient to a particular treatment, or the number of patients might be determined randomly, or the treatment–assignment probabilities might be unequal for some or all patients, or the treatment-assignment probabilities might be zero for particular treatment–patient combinations. Randomization tests can easily accommodate these circumstances by mimicking the randomization procedure in the derivation of the reference distribution (Edgington & Onghena 2007). Randomization tests can even be applied to test for treatment effects in randomized N-of-1 clinical trials, which, by definition, employ unconventional randomization procedures (see Dugard et al. 2012; Edgington 1967; Ferron & Levin 2014; Onghena 1994, 2007, 2016; Onghena & Edgington 2005).

Another distinctive feature of randomization tests is that they can be used with any customized test statistic. Other statistical tests have to use a specific test statistic, for example, the t, the F, or the χ^2 statistic, for which the small-sample or asymptotic sampling distribution has been derived mathematically. For a randomization test, however, the reference distribution of any test statistic is derived specifically for each data set, by combining the response vector and the design vector. The test statistic is chosen freely to fit the purpose of the research, and this makes the randomization test a very versatile inferential technique for a multiplicity of research contexts (Berger 2000; Dugard 2014). Because of this flexibility and versatility, the randomization test can be considered a generic procedure to derive a reference distribution rather than a narrowly defined statistical test. For example, a t test statistic can be used in a randomization test (to obtain a so-called randomization t test), but we can also use the difference between medians or the variance ratio as test statistics if those statistics more closely fit the researcher's hypothesis (Edgington & Onghena 2007).

13.2.6.3 No Random Sampling and No Distributional Assumptions Required

The most prominent feature of a randomization test is that no random sampling is assumed. Consequently, in research situations in which random sampling is impractical or impossible, a randomization test is best suited (Edgington 1966; Hunter & May 2003; Ludbrook & Dudley 1998). For example, in clinical trials, it is usually not realistic to assume that the patients have been sampled from a hypothetical infinitely large population; patients enter the trial one by one based on hospital admissions and trial selection criteria, and constitute a convenience sample rather than a random sample (Berger 2000; Cleophas et al. 2009; Onghena & Edgington 2005).

Furthermore, because no random sampling is assumed, there is no need for a population model, and consequently, there is also no need for assumptions about population distributions. The randomization test is defined within a so-called randomization model that only needs the observed data, a randomized design, and a sensitive test statistic (Keller 2012; Kempthorne 1955; Lehmann 1975; Lehmann & Romano 2005; Pitman, 1937a,b, 1938). In the same vein, notice that the null hypothesis in Equation 13.7 is not formulated in terms of population parameters, but that it is formulated at the level of the observed responses (Edgington & Onghena 2007).

13.2.6.4 Frequentist, Conditional, and Causal Inference

A final characteristic of the randomization test concerns the nature of the statistical inference that is achieved. First of all, this inference is frequentist (Bandyopadhyay & Forster 2011; Wasserman 2008; Zelen 1998). The randomization test fits within the frequentist inferential frameworks of Fisher and Neyman–Pearson. The p value in Equation 13.12 or Equation 13.14 can be interpreted directly in frequentist terms: This p value is the relative frequency of reference statistics that are as extreme as, or more extreme than, the observed statistic in the reference distribution. In other words, it is the relative frequency of trials that are at the same distance or even more distant to the null hypothesis than the observed trial if an infinite number of repetitions of the trial for a true null hypothesis is conceived. Furthermore, the operating characteristics of the randomization test can be expressed as long-term first and second kind error rates (Hinkelmann & Kempthorne 2005, 2008, 2012; Lehmann 1975).

Second, the inference from a randomization test is conditional on the observed data, just like Fisher's Exact Test is conditional on the marginal totals (Hothorn et al. 2006; Krauth 1988; Pesarin & Salmaso 2010). In fact, in a randomized trial, Fisher's Exact Test is a randomization test for comparing two treatments with a dichotomous outcome variable (Agresti 2013; Edgington & Onghena 2007; Mehta 1994). This conditionality is obvious from the definition of the test statistic, $s_k = s(x_k, y)$, as a function s of a random design vector and a fixed response vector (no subscript k for vector y).

Last and most importantly, the inference from a randomization test is causal. The null hypothesis of no treatment effect can be interpreted as a null hypothesis about the absence of a causal relation. The randomization test itself focuses on the functional relation between the manipulated factor X and the outcome variable Y, and derives a p value by counterfactual reasoning (Holland 2005; Lewis 1973a,b). Suppose that X has only two levels (treatment T and control C), then the causal effect of X on outcome Y in a clinical trial is defined as the difference in Y between T and C for a given patient. The average causal effect is the average difference over all patients. The fundamental problem of causal inference is that it is impossible in a between-subjects group comparison study to observe a patient in T and C simultaneously. One of the outcomes is observed, and the other one is missing. The randomization test offers an elegant solution to this fundamental problem: If the null hypothesis is true, we know both outcomes because the outcome Y in T is identical to the outcome Y in C (Holland 1986; Imbens & Rubin 2015; Rubin 1974).

13.3 Two Sides of the Same Coin

After this presentation of the randomization test, we can more fully appreciate the link between randomization and the randomization test. This link is obvious when looking closely at Fisher's famous quote about the conclusion we can draw from a significance test:

> The force with which such a conclusion is supported is logically that of a simple disjunction: Either an exceptionally rare chance has occurred, or the theory of random distribution is not true (Fisher 1956, p. 39).

In this quote we can interpret the "conclusion" as the "rejection of the null hypothesis" and "the theory of random distribution" as "the null hypothesis." Thus, Fisher (1956) explicitly conceptualized the conclusion drawn from a significance test as an exclusive disjunction: If we obtain a small p value, then either we witnessed an improbable event or the null hypothesis is not true. The improbable event happens in a clinical trial if the patients are assigned in such a way that, "by accident," more favorable basic values end up in the treatment condition and more unfavorable ones end up in the control condition, erroneously suggesting a treatment effect that is nonexistent. In the hypothetical example of testing a new diet against a control, this would mean that the new diet has no effect but that, "by accident," patients that have less weight have been assigned to the treatment condition. In Table 13.1, this happens in Assignment 1, with patients weighing 62 and 75 kg assigned to treatment and patients weighing 76 and 127 kg assigned to control. If these weights were the observed responses in a randomization test, then the one-tailed p value would be $1/K = 0.125$. In virtual repetitions of this trial, this event is expected to happen 12.5% of the time. By setting the level of significance α at a small amount, we can control the first kind error rate at a small and known value.

In sum, the randomization test derives its statistical validity by virtue of the randomization as actually used in the clinical trial. Conversely, if a randomized design is used, it makes sense to complement the trial and measurement of the outcomes with a calculation of the randomization test p value because it provides a quantification of the probability of these outcomes if the null hypothesis is true. In this way, randomization and the randomization test are intimately linked: They are two sides of the same coin.

References

Agresti, A. (2013). *Categorical Data Analysis* (3rd ed.). Hoboken, NJ: Wiley.

Bandyopadhyay, P. S., & Forster, M. R. (Eds.) (2011). *Handbook of the Philosophy of Science, Vol. 7: Philosophy of Statistics.* Amsterdam, The Netherlands: Elsevier.

Basu, D. (1975). Statistical information and likelihood [with Discussion]. *Sankhyā: The Indian Journal of Statistics, Series A, 37,* 1–71.

Basu, D. (1980). Randomization analysis of experimental data: The Fisher randomization test. *Journal of the American Statistical Association, 75,* 575–582.

Begg, C. B. (2015). Ethical concerns about adaptive randomization. *Clinical Trials, 12,* 101.

Begg, C., Cho, M., Eastwood, S., Horton, R., Moher, D., Olkin, I., Pitkin, R., Rennie, D., Schulz, K. F., Simel, D., & Stroup, D. F. (1996). Improving the quality of reporting of randomized controlled trials: The CONSORT statement. *Journal of the American Medical Association, 276,* 637–639.

Berger, V. W. (2000). Pros and cons of permutation tests in clinical trials. *Statistics in Medicine, 19,* 1319–1328.

Berger, V. W. (2004). On the generation and ownership of alpha in medical studies. *Controlled Clinical Trials, 25,* 613–619.

Berger, V. W. (2007). Drawbacks to non-integer scoring for ordered categorical data. *Biometrics, 63,* 298–299.

Berger, V. W. (2008). Letter to the editor. *Biometrical Journal, 50,* 1–3.

Berger, V. W., & Bears, J. D. (2003). When can a clinical trial be called 'randomized'? *Vaccine, 21,* 468–472.

Berry, K. J., Johnston, J. E., & Mielke, P. W. (2014). *A Chronicle of Permutation Statistical Methods: 1920–2000, and Beyond.* New York: Springer.

Berry, S., Carlin, B., Lee, J., & Müller, P. (2010). *Bayesian Adaptive Methods for Clinical Trials.* Boca Raton, FL: CRC Press.

Campbell, D. T., & Stanley, J. C. (1963). *Experimental and Quasi-Experimental Designs for Research.* Chicago: Rand McNally.

Cleophas, T. J., Zwinderman, A. H., Cleophas, T. F., & Cleophas, E. P. (2009). Clinical trials do not use random samples anymore. In Cleophas, T. J., Zwinderman, A. H., Cleophas, T. F., & Cleophas, E. P. (Eds.), *Statistics Applied to Clinical Trials* (4th ed.) (pp. 367–374). New York: Springer.

Cochran, W., & Cox, G. (1950). *Experimental Designs.* New York: Wiley.

Cochrane, A. L. (1972). *Effectiveness and Efficiency: Random Reflections on Health Services.* London, UK: Nuffield Provincial Hospitals Trust.

Cook, T. D., & Campbell, D. T. (1979). *Quasi-experimentation: Design and Analysis Issues for Field Settings.* Chicago: Rand McNally.

Cox, D. R., & Hinkley, D. V. (1974). *Theoretical Statistics.* London, UK: Chapman and Hall.

Cox, D. R., & Reid, N. (2000). *The Theory of the Design of Experiments.* Boca Raton, FL: Chapman & Hall/CRC.

David, H. A. (1995). First (?) occurrences of common terms in mathematical statistics. *The American Statistician, 49,* 121–133.

David, H. A. (2008). The beginnings of randomization tests. *The American Statistician, 62,* 70–72.

Dugard, P. (2014). Randomization tests: A new gold standard? *Journal of Contextual Behavioral Science, 3,* 65–68.

Dugard, P., File, P., & Todman, J. (2012). *Single-Case and Small-n Experimental Designs: A Practical Guide to Randomization Tests* (2nd ed.). New York: Routledge Academic.

Edgington, E. S. (1966). Statistical inference and nonrandom samples. *Psychological Bulletin, 66,* 485–487.

Edgington, E. S. (1967). Statistical inference from $N = 1$ experiments. *The Journal of Psychology, 65,* 195–199.

Edgington, E. S. (1970). Hypothesis testing without fixed levels of significance. *The Journal of Psychology, 76,* 109–115.

Edgington, E. S. (1986). Randomization tests. In S. Kotz & N. L. Johnson (Eds.), *Encyclopedia of Statistical Sciences, Vol. 7* (pp. 530–538). New York: Wiley.

Edgington, E. S., & Onghena, P. (2007). *Randomization Tests* (4th ed.). Boca Raton, FL: Chapman & Hall/CRC.

Efron, B., & Tibshirani, R. J. (1993). *An Introduction to the Bootstrap.* New York: Chapman & Hall.

Ferron, J. M., & Foster-Johnson, L. (1998). Analyzing single-case data with visually guided randomization tests. *Behavior Research Methods, Instruments, & Computers, 30,* 698–706.

Ferron, J. M., & Jones, P. K. (2006). Tests for the visual analysis of response-guided multiple baseline data. *The Journal of Experimental Education, 75,* 66–81.

Ferron, J. M., & Levin, J. R. (2014). Single-case permutation and randomization statistical tests: Present status, promising new developments. In T. R. Kratochwill & J. R. Levin (Eds.), *Single-Case Intervention Research: Statistical and Methodological Advances* (pp. 153–183). Washington, DC: American Psychological Association.

Finch, P. D. (1986). Randomization—I. In S. Kotz & N. L. Johnson (Eds.), *Encyclopedia of Statistical Sciences, Vol. 7* (pp. 516–519). New York: Wiley.

Fisher, R. A. (1925). *Statistical Methods for Research Workers.* London, UK: Oliver & Boyd.

Fisher, R. A. (1926). The arrangement of field experiments. *Journal of the Ministry of Agriculture of Great Britain, 33,* 503–513.

Fisher, R. A. (1935). *The Design of Experiments.* London, UK: Oliver & Boyd.

Fisher, R. A. (1956). *Statistical Methods and Scientific Inference.* London, UK: Oliver & Boyd.

Folks, J. L. (1984). Use of randomization in experimental research. In K. Hinkelmann (Ed.), *Experimental Design, Statistical Models, and Genetic Statistics: Essays in Honor of Oscar Kempthorne* (pp. 17–32). New York: Marcel Dekker.

Freedman, B. (1987). Equipoise and the ethics of clinical research. *The New England Journal of Medicine, 317,* 141–145.

Gabriel, K. R., & Hall, W. J. (1983). Rerandomization inference on regression and shift effects: Computationally feasible methods. *Journal of the American Statistical Association, 78,* 827–836.

Gabriel, K. R., & Hsu, C.-F. (1983). Evaluation of the power of rerandomization tests, with application to weather modification experiments. *Journal of the American Statistical Association, 78,* 766–775.

Good, P. I. (2005). *Permutation, Parametric and Bootstrap Tests of Hypotheses* (3rd ed.). New York: Springer.

Greenberg, B. G. (1951). Why randomize? *Biometrics, 7,* 309–322.

Hacking, I. (1988). Telepathy: Origins of randomization in experimental design. *Isis, 79,* 427–451.

Higgins, J. P. T., & Green, S. (Eds.) (2011). *Cochrane Handbook for Systematic Reviews of Interventions: Version 5.1.0* [updated March 2011]. The Cochrane Collaboration. Available from http://handbook.cochrane.org

Hinkelmann, K., & Kempthorne, O. (2005). *Design and Analysis of Experiments, Vol. 2: Advanced Experimental Design.* Hoboken, NJ: Wiley.

Hinkelmann, K., & Kempthorne, O. (2008). *Design and Analysis of Experiments, Vol. 1: Introduction to Experimental Design* (2nd ed.). Hoboken, NJ: Wiley.

Hinkelmann, K., & Kempthorne, O. (2012). *Design and Analysis of Experiments, Vol. 3: Special Designs and Applications.* Hoboken, NJ: Wiley.

Hoeffding, W. (1952). The large sample power of tests based on permutations of observations. *Annals of Mathematical Statistics, 23,* 169–192.

Holland, P. W. (1986). Statistics and causal inference. *Journal of the American Statistical Association, 81,* 945–960.

Holland, P. W. (2005). Counterfactual reasoning. In B. S. Everitt & D. C. Howell (Eds.), *Encyclopedia of Statistics in Behavioral Science, Vol. 1* (pp. 420–422). Chichester, UK: Wiley.

Hollander, M., Wolfe, D. A., & Chicken, E. (2014). *Nonparametric Statistical Methods* (3rd ed.). Hoboken, NJ: Wiley.

Hothorn, T., Hornik, K., van de Wiel, M. A., & Zeileis, A. (2006). A Lego system for conditional inference. *The American Statistician, 60,* 257–263.

Hu, F., & Rosenberger, W. (2006). *The Theory of Response-Adaptive Randomization in Clinical Trials.* Hoboken NJ: Wiley-Interscience.

Hunter, M. A., & May, R. B. (2003). Statistical testing and null distributions: What to do when samples are not random. *Canadian Journal of Experimental Psychology/ Revue Canadienne de Psychologie Expérimentale, 57,* 176–188.

Imbens, G. W., & Rubin, D. B. (2015). *Causal Inference in Statistics, Social, and Biomedical Sciences: An Introduction.* New York: Cambridge University Press.

Jadad, A. R., & Enkin, M. W. (2007). *Randomized Controlled Trials: Questions, Answers and Musings* (2nd ed.). Oxford, UK: Blackwell.

Kadane, J. B., & Seidenfeld, T. (1990). Randomization in a Bayesian perspective. *Journal of Statistical Planning and Inference, 25,* 329–345.

Keller, B. (2012). Detecting treatment effects with small samples: The power of some tests under the randomization model. *Psychometrika, 77,* 324–338.

Kempthorne, O. (1952). *The Design and Analysis of Experiments.* New York: Wiley.

Kempthorne, O. (1955). The randomization theory of experimental inference. *Journal of the American Statistical Association, 50,* 946–967.

Kempthorne, O. (1977). Why randomize? *Journal of Statistical Planning and Inference, 1,* 1–25.

Kempthorne, O. (1986). Randomization—II. In S. Kotz & N. L. Johnson (Eds.), *Encyclopedia of Statistical Sciences, Vol. 7* (pp. 519–524). New York: Wiley.

Kempthorne, O., & Doerfler, T. E. (1969). The behavior of some significance tests under experimental randomization. *Biometrika, 56,* 231–248.

Kempthorne, O., & Folks, J. L. (1971). *Probability, Statistics, and Data Analysis.* Ames, IA: Iowa State University Press.

Krauth, J. (1988). *Distribution-Free Statistics: An Application-Oriented Approach.* New York: Elsevier.

Lehmann, E. L. (1975). *Nonparametrics: Statistical Methods Based on Ranks.* San Francisco: Holden-Day.

Lehmann, E. L., & Romano, J. P. (2005). *Testing Statistical Hypotheses* (3rd ed.). New York: Springer.

Lewis, D. (1973a). Causation. *Journal of Philosophy, 70,* 556–567.

Lewis, D. (1973b). *Counterfactuals.* Oxford: Blackwell.

Lindley, D. V. (1982). The role of randomization in inference. *Proceedings of the Biennial Meeting of the Philosophy of Science Association, 2,* 431–446.

Ludbrook, J., & Dudley, H. (1998). Why permutation tests are superior to *t* and *F* tests in biomedical research. *The American Statistician, 52,* 127–132.

Manly, B. F.J. (2007). *Randomization, Bootstrap and Monte Carlo Methods in Biology* (3rd ed.). Boca Raton, FL: Chapman & Hall/CRC.

Matthews, J. R. (this volume). Randomization and bias in historical perspective.

Mehta, C. R. (1994). The exact analysis of contingency tables in medical research. *Statistical Methods in Medical Research, 3,* 135–156.

Onghena, P. (1994). The power of randomization tests for single-case designs. Unpublished doctoral dissertation, Department of Psychology, Katholieke Universiteit Leuven, Belgium.

Onghena, P. (2007). *N-of-1 Randomized Clinical Trials.* In The Biomedical & Life Sciences Collection, Henry Stewart Talks Ltd, London (online at https:// hstalks.com/t/555/)

Onghena, P. (2016). *Randomization in N-of-1 clinical trials: Is It Possible to Draw Causal Inferences from Single-Patient Data?* In The Biomedical & Life Sciences Collection, Henry Stewart Talks Ltd, London (online at https://hstalks.com/bs/3311/).

Onghena, P., & Edgington, E. S. (2005). Customization of pain treatments: Single-case design and analysis. *Clinical Journal of Pain, 21,* 56–68.

Peirce, C. S., & Jastrow, J. (1885). On small differences in sensation. *Memoirs of the National Academy of Sciences, 3,* 73–83.

Pesarin, F., & Salmaso, L. (2010). *Permutation Tests for Complex Data: Theory, Applications and Software.* Chichester, UK: Wiley.

Pitman, E. J. G. (1937a). Significance tests which may be applied to samples from any populations. *Journal of the Royal Statistical Society Series B, 4,* 119–130.

Pitman, E. J. G. (1937b). Significance tests which may be applied to samples from any populations. II: The correlation coefficient. *Journal of the Royal Statistical Society Series B, 4,* 225–232.

Pitman, E. J. G. (1938). Significance tests which may be applied to samples from any populations. III: The analysis of variance test. *Biometrika, 29,* 322–335.

Rosenberger, W. F., & Lachin, J. M. (2015). *Randomization in Clinical Trials: Theory and Practice* (2nd ed.). Hoboken, NJ: Wiley.

Rubin, D. (1974). Estimating causal effects of treatments in randomized and nonrandomized studies. *Journal of Educational Psychology, 66,* 688–701.

Rubin, D. B. (1978). Bayesian inference for causal effects: The role of randomization. *The Annals of Statistics, 6,* 34–58.

Rubin, D. B. (2008). For objective causal inference, design trumps analysis. *The Annals of Applied Statistics, 2,* 808–840.

Saint-Mont, U. (2015) Randomization does not help much, comparability does. *PLoS ONE, 10*(7). e0132102. doi:10.1371/journal.pone.0132102

Salsburg, D. (2001). *The Lady Tasting Tea: How Statistics Revolutionized Science in the Twentieth Century.* New York: W. H. Freeman.

Savage, L. J. (1962). Subjective probability and statistical practice. In G. A. Barnard & D. R. Cox (Eds.), *The Foundations of Statistical Inference: A Discussion* (pp. 9–35). London, UK: Methuen.

Senn, S. (1994). Fisher's game with the devil. *Statistics in Medicine, 13,* 217–230.

Shadish, W. R., Cook, T. D., & Campbell, D. T. (2002). *Experimental and Quasi-experimental Designs for Generalized Causal Inference.* Boston: Houghton Mifflin.

Stigler, S. M. (1978). Mathematical statistics in the early states. *Annals of Statistics, 6,* 239–265.

Stigler, S. M. (1986). *The History of Statistics: The Measurement of Uncertainty before 1900.* Cambridge, MA: Harvard University Press.

Stone, M. (1969). The role of experimental randomization in Bayesian statistics: Finite sampling and two Bayesians. *Biometrika, 56,* 681–683.

Wasserman, L. (2008). Comment on article by Gelman. *Bayesian Analysis, 3,* 463–466.

Wasserstein, R. L., & Lazar, N. A. (2016). The ASA's statement on *p*-values: Context, process, and purpose. *The American Statistician, 70,* 129–133.

Zelen, M. (1998). Inference. In P. Armitage & T. Colton (Eds.), *Encyclopedia of Biostatistics* (pp. 2035–2046). New York: Wiley.

14

Randomization Tests or Permutation Tests?
A Historical and Terminological Clarification

Patrick Onghena

CONTENTS

14.1 Introduction...209
14.2 Historical Clarification..210
    14.2.1 Fisher (1935) 210
    14.2.2 Pitman (1937a,b, 1938) and Welch (1937) 212
    14.2.3 Stigler's Law of Eponymy...214
14.3 Terminological Clarification...216
    14.3.1 Randomization Tests and Permutation Tests............................217
    14.3.2 Randomization Tests and Computer Intensive Tests 218
    14.3.3 Randomization Tests and Exact Tests...219
    14.3.4 Randomization Tests and Exchangeability...............................219
14.4 Conclusion 220
References...221

14.1 Introduction

Randomization tests, as defined in Chapter 13 ("Randomization and the Randomization Test: Two Sides of the Same Coin"), are sometimes labeled "permutation tests." Furthermore, they tend to have Fisher's name attached to them, resulting in terms such as "Fisher's randomization test" (Basu 1980; Jacquez & Jacquez 2002) or "Fisher's permutation test" (Oden & Wedel 1975; Soms 1977). In this chapter, we will provide a clarification of both the history and the current use of these terms. In the historical clarification, we will argue that statisticians other than Fisher played a more decisive role in the development of a coherent theory of the randomization model. In the terminological clarification, we will emphasize that there is a crucial difference between randomization tests and permutation tests. In addition, we will discuss the relationship of randomization tests to other computer-intensive tests and exact tests, and the (ir)relevance of the concept of "exchangeability" for the validity of randomization tests and permutation tests.

14.2 Historical Clarification

The history of the relatively young scientific discipline of statistics is filled with intriguing scientific puzzles, debates, and controversy (Box 1978; Hald 1998, 2007; Reid 1982; Stigler 1978, 1986). Current practice can still be informed and inspired by the origins of these puzzles and the genesis of their solutions. In this section, we want to revisit Fisher (1935), Pitman (1937a,b, 1938), and Welch (1937) regarding the development of the randomization model for statistical inference. We will argue that Stigler's law of eponymy also applies for the randomization test: No scientific discovery is named after its original discoverer (Gieryn 1980).

14.2.1 Fisher (1935)

Fisher (1935) allegedly introduced the randomization test in the third chapter of *The Design of Experiments* in which he re-analyzed Charles Darwin's data on the difference in height between cross- and self-fertilized corn (Alf & Abrahams 1972; Basu 1980; Box & Andersen 1955; Jacquez & Jacquez 2002; Manly 2007). However, after closer examination of that chapter, it is obvious that Fisher (1935) did not embrace the randomization model at all. In every section of that chapter, Fisher (1935) firmly endorsed the classical random sampling model:

> The most general statement of our null hypothesis is therefore that the limits to which these two averages tend are equal. *The theory of errors* enables us to test a somewhat more limited hypothesis, which, by wide experience, has been found to be appropriate to the metrical characters of experimental material in biology. The disturbing causes which introduce discrepancies in the means of measurements of similar material are found to produce quantitative effects which conform satisfactorily to a theoretical distribution known as *the normal law of frequency of error*. It is this circumstance that makes it appropriate to choose, as the null hypothesis to be tested, one for which an *exact* statistical criterion is available, namely that the two groups of measurements are *samples drawn from the same normal population*. On the basis of this hypothesis we may proceed to compare the average difference in height, between the crossfertilised and the self-fertilised plants, with such differences as might be expected between these averages, in view of the observed discrepancies between the heights of plants of like origin (pp. 39–40, italics added).

Thus, from the onset of this chapter, Fisher (1935) was clear about his goal and expressed strong support for "the theory of errors" a.k.a. "the normal law of frequency of error." This enabled him to use the results of "Student's" *t* test to contest Galton's analysis of Darwin's data:

It has been mentioned that "Student's" *t* test, in conformity with the classical theory of errors, is appropriate to the null hypothesis that the two groups of measurements are *samples drawn from the same normally distributed population*. This is the type of null hypothesis which experimenters, *rightly in the author's opinion*, usually consider it appropriate to test, for reasons not only of practical convenience, but *because the unique properties of the normal distribution make it alone suitable for general application* (p. 50, italics added).

It is only to ward off critics of "Student's" *t* test regarding the assumption of the normally distributed population that Fisher (1935) used the randomization argument in the last few pages of that chapter. However, he did not loosen the random sampling assumption:

On the hypothesis that the two series of seeds are random samples from identical populations, and that their sites have been assigned to members of each pair independently at random, the 15 differences of Table 3 would each have occurred with equal frequency with a positive or with a negative sign (Fisher 1935, p. 51).

Furthermore, he left no doubt that he considered the randomization argument merely as an auxiliary device to validate the tests based on "the classical theory of errors":

The arithmetical procedure of such an examination [i.e., the randomization test] is tedious, and we shall only give the results of its application in order to show the possibility of an independent check on the more expeditious methods in common use ["Student's" *t* test] (Fisher 1935, p. 51).

He makes this point more explicit in a 1936 paper:

Actually, the statistician does not carry out this very simple and very tedious process [i.e., the randomization test], but his conclusions have no justification beyond the fact that they agree with those which could have been arrived at by this elementary method.

For the seventh edition of *The Design of Experiments*, Fisher (1960) appended a short section in the third chapter in which he curbed any enthusiasm regarding "this elementary method." In this additional section, he simultaneously claimed to have had the scoop and emphatically dismissed the "tests using the physical act of randomization" as only deserving a secondary role:

In recent years tests using the physical act of randomisation to supply (on the Null Hypothesis) a frequency distribution, have been largely advocated under the name of "Non-parametric" tests. Somewhat extravagant claims have often been made on their behalf. The example of this Chapter, published in 1935, was by many years the first of its class. The reader will realise that it was in no sense put forward to supersede the common and expeditious tests based on the Gaussian theory of errors. The utility of such nonparametric tests consists in their being able to supply confirmation whenever, rightly or, more often, wrongly, it is suspected that the simpler tests have been appreciably injured by departures from normality (p. 48).

However, Fisher (1935) cannot be given the scoop for discovering the randomization test because he never abandoned the random sampling idea. In his description of the test on page 51 of *The Design of Experiments,* he makes no conceptual difference between random sampling and random assignment. Instead, the reader is given the impression that any random process validates the use of the Gaussian theory of errors. He thought of "tests using the physical act of randomization" merely as a dispensable tool to confirm the universal validity of classical parametric tests based on random sampling.

At most, the third chapter of *The Design of Experiments* can be considered as an intermediate step in an ongoing discussion of randomization both as a design principle and as a way to validate statistical tests. Considered in Fisher's body of work, it is a provisional argument in a larger project with a different goal (parametric likelihood theory), and Fisher never returned to this argument in his statistical work ever again (Kempthorne 1976; Rubin 2005). Furthermore, as noticed by David (2008), Fisher's justification of the Gaussian theory of errors by randomization is predated by a paper of Eden and Yates (1933), prepared at Rothamsted shortly before Fisher left for University College London. Eden and Yates (1933) acknowledge Fisher's "interest and advice," but Fisher (1935) makes no reference to Eden and Yates (1933). As mentioned by Rubin (1990), Imbens and Rubin (2015), Berry et al. (2014, 2016), Fisher's ideas about randomization and the repeated-sampling evaluations of randomization distributions were predated by many other publications, most notably by Neyman (1923), "Student" (1923), and Geary (1927).

14.2.2 Pitman (1937a,b, 1938) and Welch (1937)

Edgington and Onghena (2007) pointed out that the first author who proposed the proper randomization model and who presented the randomization test as a test for its own sake was Pitman (1937a,b, 1938). Edwin J. G. Pitman was professor of Mathematics, working at the remote University of Tasmania in Australia. He "strayed into statistics" after meeting an experimenter at the State Department of Agriculture, R. A. Scott, who brought him some data and statistical analyses from field trials on potatoes, together with a copy of Fisher's *Statistical Methods for Research Workers* (Pitman 1982; Williams 1994). Pitman started studying Fisher's publications and verifying his calculations, and immediately understood the broader application of Fisher's validation argument to "significance tests which may be applied to samples from any populations" (Sprent 1994; Williams 1994). In a series of papers, he introduced the randomization test for comparing the averages of two independent groups (Pitman 1937a), the randomization test for the correlation coefficient (Pitman 1937b), and the randomization test for analysis of variance (Pitman 1938). He showed how significance tests, confidence intervals, and their approximations can be derived under a random sampling model, but he also provided demonstrations for situations in which "the two samples together form the whole population," which means that there is only random assignment and no population other than the sample itself:

> But the essential point of the method is that we do not have to worry about the populations which we do not know, but only about the sample values which we do know (Pitman 1937a, p. 129).

> In particular, the theorem is true when the two samples together form the whole population; it then needs no proof (Pitman 1937a, p. 129).

However, Pitman (1937a) made the issue of precedence complicated by humbly giving most of the credit to Fisher (1935):

> The main idea is not new, it seems to be implicit in all Fisher's writings; [Footnote: See, for example, R. A. Fisher, *The Design of Experiments*, p. 50 (Oliver and Boyd).] but perhaps the approach to the subject, frankly starting from the sample and working towards the population instead of the reverse, may be a bit of a novelty (Pitman 1937a, p. 119).

Later in his career, Pitman regretted that he wrote that disclaimer:

> I was always dissatisfied with the sentence that I wrote... I wanted to say that I really was doing something new (Pitman in a letter dated June 1986, quoted in Edgington 1995, p. 18).

The least one can say, is that it was an exaggeration to refer to "all Fisher's writings."

At about the same time and presumably independent from Pitman, Welch (1937) proposed randomization tests for Randomized Blocks and Latin Square experiments. Bernard L. Welch was working closer to the epicenter of the world's leading statistical research and development: at the University College London, where Egon Pearson was head of the Department of Statistics and Fisher head of the Department of Eugenics (Box 1978; Mardia 1990). Later in his career, he would develop the famous Welch t test (as an alternative to Student's t test without assuming equal variances, Welch 1947), but in his 1937 paper, he developed randomization tests along the same lines as Pitman did, and emphasized that the statistical inference resulting from a randomization test was not intended to go beyond the data at hand:

> In experiments in which randomization is performed, the actual arrange-
> ment of treatments on the field is one chosen at random from a predeter-
> mined set of possible arrangements. In the present paper investigation
> has been made for Randomized Blocks and Latin Square experiments,
> into the distribution of the statistic z, generated by the application to
> the observed plot yields of the whole fundamental set of arrangements,
> assuming as true the "null" hypothesis that the treatments have no dif-
> ferential effect on the plot (Welch 1937, p. 47).

> We may make, from the yields of the experiment, a *statistical* inference
> only about the situation on the particular field of the experiment, e.g. as
> in the present paper, we may be using our *statistical* method only to test
> whether all the treatments would have given identical yields on each
> plot of this particular field (p. 47, italics in original).

Welch (1937) did not refer to Pitman (1937a,b) but Pitman (1938) referred
to Welch (1937). This is remarkable because, as pointed out by Sprent (1994),
in the pre-airmail era, it took about three months to send letters and papers
between London and Hobart. Pitman graciously acknowledged Welch's
work in the introduction to his 1938 paper:

> As one of a series, this paper was planned many months ago; but it was
> not written until June of this year, 1937. It arrived in England just about
> the time when B. L. Welch's paper on the analysis of variance appeared.
> While some of its results are anticipated by Welch, the present paper
> goes deeper into the randomization theory of the simplest type of analy-
> sis of variance test (Pitman 1938, p. 322).

In the final section, he apologetically added:

> Only the simplest type of analysis of variance test has been discussed in
> this paper. I had intended another paper to follow, which would deal in
> the same way with the Latin square arrangement; but this has been dealt
> with by Welch (5). I may add that Welch's equation (49) on p. 41, giving
> the variance of W for Latin square, agrees with my own result, which
> was reached by a route quite different from his. In view of the rather
> heavy algebra involved it seems worth while publishing this confirma-
> tion of Welch's result (Pitman 1938, p. 335).

14.2.3 Stigler's Law of Eponymy

Hence, perhaps, it is no surprise that Fisher's name became attached to the
test: Pitman (1937a) credited Fisher (1935), and Welch (1937) did not refer to
Pitman (1937a,b, 1938) because they were working on opposite sides of the
globe, and in the 1930s, only snail mail was available. Furthermore, Fisher
was a fervent advocate of randomization, and in his influential publications

from a decade earlier, he promoted randomization as one of the basic principles of experimental design, together with blocking and replication (Fisher 1925, 1926).

Interestingly, another statistical heavyweight, Jerzy Neyman, did something very similar with respect to giving credit to Fisher for the randomization design principle as Edward Pitman did with respect to giving credit to Fisher for developing the test. In his notorious paper, read before the Industrial and Agricultural Research Section of the Royal Statistical Society, Neyman (1935, p. 109) stated:

> Owing to the work of R. A. Fisher, "Student" and their followers, it is hardly possible to add anything essential to the present knowledge concerning local experiments.... One of the most important achievements of the English School is their method of planning field experiments known as the method of Randomized Blocks and Latin Squares.

A few pages later, he added:

> The difficulty has been overcome by the device proposed by R. A. Fisher, which consists in making the η's random variables with mean equal to zero. For this purpose the plots within each block are randomly distributed among the different objects (Neyman 1935, p. 112).

Notwithstanding his own 1923 paper on the completely randomized design, Neyman firmly ascribed the introduction of randomization as a physical act, and later as a basis for analysis, to Fisher (Speed 1990). Neyman's biographer, Constance Reid (1982), recalls:

> On one occasion, when someone perceived him as anticipating the English statistician R. A. Fisher in the use of randomization, he objected strenuously: "…. I treated theoretically an unrestrictedly randomized agricultural experiment and the randomization was considered as a prerequisite to probabilistic treatment of the results. This is not the same as the recognition that without randomization an experiment has little value irrespective of the subsequent treatment. The latter point is due to Fisher, and I consider it as one of the most valuable of Fisher's achievements" (p. 45).

With all these credits, it might be obvious to also name the test after Fisher. In the end, he is probably one of the most influential and prolific statisticians of the twentieth century (Hald 1998, 2007; Heyde & Seneta 2001; Lehmann 2011; Rao 1992; Salsburg 2001), so his name may act as a magnet for statistical procedures that are looking for justification, recognition, and respect. In this sense, Fisher's celebrity enjoys the benefit of the Matthew effect as described in the sociology of science (Merton 1968). Stephen Stigler derived Stigler's law of eponymy from this effect: "No scientific discovery is named after its original discoverer." With Robert Merton (1968) as the true discoverer, this "law" confirms itself (Gieryn 1980).

For completeness' sake, it should be added that some authors rightly use Pitman's name as the descriptor for the test. For example, references can be found to the "Fisher–Pitman test" (Wilks 1962), the "Pitman–Welch test" (Feinstein 1985), and the "Pitman test" (Gibbons 1986b; Pratt & Gibbons 1981). These exceptions confirm the rule but usually only refer to a specific design and a specific model. For example, Wilks (1962) used the term "Fisher–Pitman test" to indicate the two-sample test in a random sampling model. Notice also that the so-called "Fisher's Exact Test" for two-by-two contingency tables is in fact a randomization test in a completely randomized design with two treatments and with two categories of responses (Edgington & Onghena 2007). The origin of Fisher's Exact Test in *The Design of Experiments* and the resulting controversies are beyond the scope of the present chapter, but the interested reader is referred to Barnard (1945, 1947, 1949), Neyman (1950), Gridgeman (1959), Berkson (1978), Basu (1980), Yates (1984), Agresti (2013), and Bi and Kuesten (2015).

Finally, it should be noted that neither Fisher (1935) nor Pitman (1937a,b, 1938) or Welch (1937) used the term "randomization test." Fisher referred to "tests using the physical act of randomization." Pitman merely referred to "significance tests which may be applied to samples from any populations," whereas Welch (1937) compared the "Normal Theory" and the "Randomization Theory." According to David (2001), the term "randomization test" makes its first appearance in Box and Andersen's 1955 paper on "Permutation Theory in the Derivation of Robust Criteria and the Study of Departures from Assumption," in which it is used as a synonym for "permutation test." However, the term "randomization test" is already prominently present in Moses's 1952 paper on "Non-parametric Statistics for Psychological Research" (the term even appears as a section title on page 133) and in Kempthorne's 1952 classical textbook, *The Design and Analysis of Experiments* (also as a section title, page 128), suggesting an even older (possibly common) source.

14.3 Terminological Clarification

As is evident from Section 14.1, the terminology surrounding "tests using the physical act of randomization" has been confusing since their first introduction. Even nowadays still all kinds of terms are used interchangeably to refer to these tests. In this chapter, we want to endorse the proposal first made by Kempthorne and Doerfler (1969) to make an explicit distinction between "randomization tests" and "permutation tests." Randomization tests are based on a random assignment model and permutation tests are based on a random sampling model. The justification of the randomization test derives from the fact that under the null hypothesis of no treatment effect, the random assignment procedure produces just a random shuffle of the responses. The justification of the permutation test derives from the fact that under the null hypothesis of identical distributions, all permutations of the responses are equally likely. This distinction is also followed by Cox and Hinkley (1974), Willmes (1987), Mewhort (2005), Zieffler et al. (2011), and Keller (2012). Randomization tests (under the strict random assignment model) are also sometimes called "rerandomization tests" (Brillinger et al. 1978; Gabriel 1979; Gabriel & Hall 1983; Gabriel & Hsu 1983; Petrondas & Gabriel 1983), as an analogy with "resampling tests," but this terminology is rare in the current scientific literature. Furthermore, this terminology might be misleading because it contains a suggestion that a new test procedure is proposed (although it is an "ordinary" randomization test). Finally, the physical act of randomization is a one-time thing, and "rerandomization" is not at stake.

14.3.1 Randomization Tests and Permutation Tests

The strict distinction between "randomization tests" and "permutation tests" based on their different justifications for their validity implies that there is no subset relation between the two types of tests, and that there is even no intersection. However, sometimes authors use the term "permutation tests" to refer to a broader category of tests including randomization tests as a special case (Ernst 2004; Good 2005; Lehmann & Romano 2005; Pesarin & Salmaso 2010; Scheffé 1959). This causes no interpretation problems in specific applied research contexts if the models and justifications are also clearly specified. However, generally speaking, having identical terms referring to different procedures and having different terms referring to identical procedures is a recipe for confusion.

Symptomatic for this confusion are the entries in the *Encyclopedia of Statistical Sciences*. In this encyclopedia, Gibbons (1986a) defined permutation tests as a particular kind of randomization test, while Edgington (1986) defined randomization tests as a particular kind of permutation test. According to Gibbons (1986a), a permutation test is a randomization test whose reference distribution is generated by permutations, and according to Edgington (1986), a randomization test is a permutation test that is based on random assignment. Such contradictory definitions can be avoided by making a clear distinction between randomization tests and permutation tests and linking them firmly to the corresponding random assignment model and random sampling model. One additional advantage of avoiding the term "permutation tests" for the broader category is that we do not have to invent another word for "permutation tests that are no randomization tests" ("permutation tests sensu stricto"). Another additional advantage of avoiding the term "permutation tests" in the definition of randomization tests is that the unfortunate reference to "permutation" as a combinatorial concept disappears. This reference is unfortunate because many random assignment schedules are not based on permutations but rather on combinations, partitions, or even on random determination of interventions points in randomized single-case phase designs (Edgington & Onghena 2007; Kratochwill & Levin 2010; Moir 1998; Onghena & Edgington 2005).

Notice that Lehmann and D'Abrera (2006) and Siegel and Castellan (1988) used the term "permutation tests" even in yet another, more restricted way. They used it to make a distinction with nonparametric rank tests. "Permutation tests," according to Lehmann and D'Abrera (2006) and Siegel and Castellan (1988), are applied to the original scores, while nonparametric rank tests are applied to the ranks. Of course, this adds to the confusion. To make things worse, in the first edition of his popular handbook, Siegel (1956) used the term "randomization tests" instead of "permutation tests" for this category of tests. In order to emphasize that the tests are applied to the original scores, other authors used the terms "component-randomization tests" or "observation-randomization tests" as opposed to "rank-randomization tests," which might be more appropriate and less confusing (Alf & Abrahams 1972; Bradley 1968; Pratt & Gibbons 1981; Wilks 1962).

14.3.2 Randomization Tests and Computer-Intensive Tests

The computations involved in performing randomization tests have much in common with other computer-intensive methods, such as the jackknife, the bootstrap, cross-validation, and Monte Carlo tests (Efron 1979; Efron & Gong 1983; Efron & Tibshirani 1991; Manly 2007; Noreen 1989; Romano 1989). This has led some authors to discuss randomization tests in a more general framework of "resampling" or "bootstrapping" (Simon & Bruce 1991; Westfall & Young 1993). In such a framework, the only difference between bootstrap significance tests and randomization tests is that with the former method the data are sampled with replacement, while in the latter, the data are sampled without replacement (Manly 2007; Romano 1989; Westfall & Young 1993).

One should, however, not overlook the fact that bootstrap significance tests and randomization tests have different theoretical foundations and different statistical properties. For example, with a randomization test, the Type I error rate is completely under control, while with a bootstrap significance test, the Type I error rate is only under control for large samples in some designs with some test statistics (Manly 2007; Noreen 1989; Rasmussen 1989; ter Braak 1992). As Efron and Tibshirani (1993) commented:

> The bootstrap distribution was originally called the "combination distribution". It was designed to extend the virtues of permutation testing to the great majority of statistical problems where there is nothing to permute. When there *is* something to permute (…) it is a good idea to do so, even if other methods like the bootstrap are also brought to bear (p. 218).

14.3.3 Randomization Tests and Exact Tests

Randomization tests and permutation tests are sometimes called "exact tests" to the effect that the terms are used as synonyms (Chernick & Friis 2003; Gacula et al. 2009; Higgins 2017; Krauth 1988). The "exactness" of the test refers to the exact control of the Type I error rate, and in this sense depends on the kind of assumptions that are involved. "Exact tests" is a term that makes sense in contrast to approximate tests or Monte Carlo tests that only control the Type I error rate approximately or asymptotically (Mehta & Patel 1983; Mehta et al. 1988, 1994; Senchaudhuri et al. 1995).

Although "exact tests" is an appealing term—who wants to use "inexact tests"?—it fails as a synonym for randomization tests or permutation tests. Also, parametric tests are exact tests if their assumptions are satisfied (see Weerahandi 1995, 2004). Furthermore, Dwass (1957) and Hope (1968) developed Monte Carlo versions of randomization tests and permutation tests, which are perfectly valid as tests on their own, but which are, by the above definition, not exact tests. These Monte Carlo versions have become increasingly popular in recent years because of their relative ease of programming as compared to the exhaustive versions in which all possibilities have to be enumerated once and only once. Notice that these Monte Carlo tests are perfectly valid if the observed test statistic is included in the reference set. By contrast, if the "exact p value" is merely estimated by a Monte Carlo procedure without including the observed test statistic in the reference set, as proposed by Senchaudhuri et al. (1995), then it is no longer a perfectly valid test on its own (Edgington & Onghena 2007; Onghena & May 1995; Phipson & Smyth 2010). If only one hypothesis is to be tested, the consequences of the miscalculation by using the wrong Monte Carlo version are usually negligible, but in a multiple testing situation, they may become substantial. As pointed out by Phipson and Smyth (2010):

> In genomic research, however, it is typically the case that many tests are to be conducted. When the number of tests is large, any systematic underestimation of p-values can lead to dangerously wrong conclusions at the family-wise level (p. 4).

14.3.4 Randomization Tests and Exchangeability

In some publications, "exchangeability" is used as a common foundational assumption of randomization tests and permutation tests (see, e.g., Commenges 2003; Good 2005; Nichols & Holmes 2002; Pesarin & Salmaso 2010; Winkler et al. 2014). Exchangeability is satisfied if the joint distribution of the n observations $y_1, \ldots y_i, \ldots y_n$ is invariant under permutations of the indices. Examples of exchangeable observations are observations from independent, identically distributed (iid) variables and observations from jointly Gaussian distributed variables with identical covariances (Draper et al. 1993; Good 2002; Greenland & Draper 2011).

Exchangeability, as a statistical concept, has its origin in Bayesian statistics, in which it is used as a "weaker" assumption than the assumption of iid variables (i.e., all iid variables are exchangeable, but not all exchangeable variables are iid) (Galambos 1986; Gelman et al. 2014; Koch 1982; Lindley & Novick 1981). Although this property of exchangeability is important for the foundation of statistical inference, its immediate relevance for the applied statistician is limited. In his discussion of a paper by Draper et al. (1993), George A. Barnard suggested the term "permutability" as a more apt term for the property of exchangeability. This suggestion reveals the *circulus in probando* if the validity of randomization tests and permutation tests is grounded in exchangeability: the observations can be permuted because they are exchangeable. This means that the observations can be permuted because the observations can be permuted. Furthermore, the validity of randomization tests is not based on the exchangeability of the observations as such, but rather on the exchangeability of all elements in the reference set. As mentioned before, many random assignment schedules do not imply permutations of the observations (Edgington & Onghena 2007; Kratochwill & Levin 2010; Moir 1998; Onghena & Edgington 2005).

For the validity of randomization tests and permutation tests, a reference to actual random assignment or actual random sampling suffices. This reference connects our statistical arithmetic to real-world phenomena and actions. Random assignment or random sampling schedules show in which way the experimental units—and therefore also the observations—are exchangeable, but the concept of "exchangeability" is not needed to provide this foundation. It is just rewording the obvious. For data collected without random assignment and without random sampling, any "observational inference" uses an *as if* modus; this means that inferences are made *as if* there was actual random assignment or actual random sampling (Dekkers 2011; Kempthorne 1979; Vandenbroucke 2004). Although the relevance of the concept of exchangeability for the validation of randomization tests and permutation tests is limited, it should be acknowledged that the concept is theoretically important, for example, to demonstrate the common foundation of Bayesian and permutational inference and the development of new statistical techniques in the absence of random assignment or random sampling (Draper et al. 1993; Good 2002; Hutson & Wilding 2012).

14.4 Conclusion

In this chapter, we clarified some terms associated with significance tests based on random assignment and we shed some light on the historical development of these tests. Clarification of terms is important for current practice and dissemination. Clarification of the historical development is important to give proper credit to the researchers involved in laying the foundations of the modern use of these statistical techniques and to understand and appreciate the basic questions and intricacies that they were focusing on.

In sum, Fisher (1935) was very successful in his advocacy for the use of randomization in experimental research and for its importance for the validity of statistical tests, but he never systematically elaborated or even showed any interest for the theory behind the significance tests based on random assignment alone. Pitman (1937a,b, 1938) and Welch (1937) picked up that gauntlet and proposed randomization tests as statistical techniques for their own sake.

For the modern use of these statistical techniques, it is crucial to acknowledge the difference between a random assignment model and a random sampling model. This is accomplished most conveniently by following the terminology of Kempthorne and Doerfler (1969): Significance tests based on the random assignment model are called "randomization tests." Other computer-intensive tests, such as permutation tests and bootstrap tests, are based on the random sampling model and eventually other accompanying assumptions. This difference is important to know exactly what kind of assumptions are involved and whether conditional or unconditional power calculations are most appropriate in the planning stage of the clinical trial (Keller 2012; Kempthorne & Doerfler 1969; Pesarin & De Martini 2002). Finally, the difference between the two kinds of models is crucial to understand the kind of inference that is aimed at (inference to a causal proposition or inference to the sampled population/generalization), and consequently to understand the kind of criteria that have to be used to judge their validity (internal vs. external validity) (Shadish et al. 2002).

For Tukey (1993), statistical inference in clinical trials based on the actual random assignment constitutes the "platinum standard" of data analysis:

> "platinum standard" describes the ultimate of tightness: probability statements that depend on only exactly how the trial was conducted— not at all on assumptions, such as shapes of probability distributions, that are less (often far less) than completely verifiable (pp. 266–267).

To recognize this platinum standard, we recommend using the right term unambiguously: "randomization test" or, if you want to pay tribute to the founding fathers: "Pitman-Welsh randomization test."

References

Agresti, A. (2013). *Categorical Data Analysis* (3rd ed.). Hoboken, NJ: Wiley.

Alf, E. F., & Abrahams, N. M. (1972). Comment on component-randomization tests. *Psychological Bulletin, 77,* 223–224.

Barnard, G. A. (1945). A new test for 2×2 tables. *Nature, 156*(3954), 177.

Barnard, G. A. (1947). Significance tests for 2×2 tables. *Biometrika, 34,* 123–138.

Barnard, G. A. (1949). Statistical inference. *Journal of the Royal Statistical Society, Series B, 11,* 115–149.

Basu, D. (1980). Randomization analysis of experimental data: The Fisher randomization test. *Journal of the American Statistical Association, 75,* 575–582.

Berkson, J. (1978). In dispraise of the exact test. *Journal of Statistic Planning and Inference, 2,* 27–42.

Berry, K. J., Johnston, J. E., & Mielke, P. W. (2014). *A Chronicle of Permutation Statistical Methods: 1920–2000, and Beyond.* New York: Springer.

Berry, K. J., Mielke, P. W., & Johnston, J. E. (2016). *Permutation Statistical Methods: An Integrated Approach.* New York: Springer.

Bi, J., & Kuesten, C. (2015). Revisiting Fisher's 'Lady Tasting Tea' from a perspective of sensory discrimination testing. *Food Quality and Preference, 43,* 47–52.

Box, G. E. P., & Andersen, S. L. (1955). Permutation theory in the derivation of robust criteria and the study of departures from assumption, *Journal of the Royal Statistical Society, Series B, 17,* 1–34.

Box, J. F. (1978). *R. A. Fisher: The Life of a Scientist.* New York: Wiley.

Bradley, J. V. (1968). *Distribution-Free Statistical Tests.* Englewood Cliffs, NJ: Prentice-Hall.

Brillinger, D. R., Jones, L. V., & Tukey, J. W. (1978). *The Management of Weather Resources II: The Role of Statistics in Weather Resources Management.* Washington, DC: U.S. Government Printing Office.

Chernick, M. R., & Friis, R. H. (2003). *Introductory Biostatistics for the Health Sciences: Modern Applications Including Bootstrap.* New York: Wiley.

Commenges, D. (2003). Transformations which preserve exchangeability and application to permutation tests. *Nonparametric Statistics, 15,* 171–185.

Cox, D. R., & Hinkley, D. V. (1974). *Theoretical Statistics.* London, UK: Chapman and Hall.

David, H. A. (2001). First (?) occurrence of common terms in statistics and probability. In H. A. David & A. W. F. Edwards (Eds.), *Annotated Readings in the History of Statistics* (pp. 219–228 and Appendix B). New York: Springer.

David, H. A. (2008). The beginnings of randomization tests. *The American Statistician, 62,* 70–72.

Dekkers, O. M. (2011). On causation in therapeutic research: Observational studies, randomized experiments and instrumental variable analysis. *Preventive Medicine, 53,* 239–241.

Draper, D., Hodges, J. S., Mallows, C. L., & Pregibon, D. (1993). Exchangeability and data analysis (with discussion). *Journal of the Royal Statistical Society, Series A, 156,* 9–28.

Dwass, M. (1957). Modified randomization tests for nonparametric hypotheses. *Annals of Mathematical Statistics, 28,* 181–187.

Eden, T., & Yates, F. (1933). On the validity of Fisher's z test when applied to an actual example of non-normal data. *The Journal of Agricultural Science, 23,* 6–17.

Edgington, E. S. (1986). Randomization tests. In S. Kotz & N. L. Johnson (Eds.), *Encyclopedia of Statistical Sciences, Vol. 7* (pp. 530–538). New York: Wiley.

Edgington, E. S. (1995). *Randomization Tests* (3rd ed.). New York: Marcel Dekker.

Edgington, E. S., & Onghena, P. (2007). *Randomization Tests* (4th ed.). Boca Raton, FL: Chapman & Hall/CRC.

Efron, B. (1979). Bootstrap methods: Another look at the jackknife. *The Annals of Statistics, 7,* 1–26.

Efron, B., & Gong, G. (1983). A leisurely look at the bootstrap, the jackknife, and cross-validation. *The American Statistician, 37,* 36–48.

Efron, B., & Tibshirani, R. J. (1991). Statistical data analysis in the computer age. *Science, 253*(5018), 390–395.

Efron, B., & Tibshirani, R. J. (1993). *An Introduction to the Bootstrap.* Boca Raton, FL: Chapman & Hall/CRC.

Ernst, M. D. (2004). Permutation methods: A basis for exact inference. *Statistical Science, 19,* 676–685.

Feinstein, A. R. (1985). *Clinical Epidemiology: The Architecture of Clinical Research.* Philadelphia, PA: W. B. Saunders Company.

Fisher, R. A. (1925). *Statistical Methods for Research Workers.* Edinburgh, UK: Oliver & Boyd.

Fisher, R. A. (1926). The arrangement of field experiments. *Journal of the Ministry of Agriculture of Great Britain, 33,* 503–513.

Fisher, R. A. (1935). *The Design of Experiments.* Edinburgh, UK: Oliver & Boyd.

Fisher, R. A. (1936). "The Coefficient of Racial Likeness" and the future of craniometry. *The Journal of the Royal Anthropological Institute of Great Britain and Ireland, 66,* 57–63.

Fisher, R. A. (1960). *The Design of Experiments* (7th ed.). Edinburgh, UK: Oliver & Boyd.

Gabriel, K. R. (1979). Some statistical issues in weather experimentation. *Communications in Statistics: Theory and Methods, 8,* 975–1015.

Gabriel, K. R., & Hall, W. J. (1983). Rerandomization inference on regression and shift effects: Computationally feasible methods. *Journal of the American Statistical Association, 78,* 827–836.

Gabriel, K. R., & Hsu, C. F. (1983). Power studies of rerandomization tests, with application to weather modification experiments. *Journal of the American Statistical Association, 78,* 766–775.

Gacula, M. C., Jr., Singh, J., Bi, J., & Altan, S. (2009). *Statistical Methods in Food and Consumer Research.* Amsterdam, The Netherlands: Elsevier.

Galambos, J. (1986). Exchangeability. In S. Kotz and N. L. Johnson (Eds.). *Encyclopedia of Statistical Sciences, Vol. 3* (2nd ed.) (pp. 2136–2140). New York: Wiley.

Geary, R. C. (1927). Some properties of correlation and regression in a limited universe. *Metron: International Journal of Statistics, 7,* 83–119.

Gelman, A., Carlin, J. B., Stern, H. S., Dunson, D. B., Vehtari, A., & Rubin, D. B. (2014). *Bayesian Data Analysis* (3rd ed.). Boca Raton, FL: Chapman & Hall/CRC.

Gibbons, J. D. (1986a). Permutation tests. In S. Kotz & N. L. Johnson (Eds.), *Encyclopedia of Statistical Sciences, Vol. 6* (p. 690). New York: Wiley.

Gibbons, J. D. (1986b). Pitman tests. In S. Kotz & N. L. Johnson (Eds.), *Encyclopedia of Statistical Sciences, Vol. 6* (pp. 740–743). New York: Wiley.

Gieryn, T. F. (Ed.) (1980). *Science and Social Structure: A festschrift for Robert K. Merton.* New York: Academy of Sciences.

Good, P. (2002). Extensions of the concept of exchangeability and their applications. *Journal of Modern Applied Statistical Methods, 1,* 243–247.

Good, P. (2005). *Permutation, Parametric and Bootstrap Tests of Hypotheses* (3rd ed.). New York: Springer.

Greenland, S., & Draper, D. (2011). Exchangeability. In M. Lovric (Ed.), *International Encyclopedia of Statistical Science* (pp. 474–476). Heidelberg, Germany: Springer.

Gridgeman, N. T. (1959). The lady tasting tea, and allied topics. *Journal of the American Statistical Association, 54*, 776–783.

Hald, A. (1998). *A History of Mathematical Statistics.* New York: Wiley.

Hald, A. (2007). *A History of Parametric Statistical Inference from Bernoulli to Fisher, 1713–1935.* New York: Springer.

Heyde, C. C., & Seneta, E. (2001). *Statisticians of the Centuries.* New York: Springer.

Higgins, J. J. (2017). *Introduction to Modern Nonparametric Statistics* [Online]. Content Technologies Inc.

Hope, A. C. A. (1968). A simplified Monte Carlo significance test procedure. *Journal of the Royal Statistical Society, Series B, 30*, 582–598.

Hutson, A. D., & Wilding, G. E. (2012). Maintaining the exchangeability assumption for a two-group permutation test in the non-randomized setting. *Journal of Applied Statistics, 39*, 1593–1603.

Imbens, G. W., & Rubin, D. B. (2015). *Causal Inference in Statistics, Social, and Biomedical Sciences: An Introduction.* New York: Cambridge University Press.

Jacquez, J. A., & Jacquez, G. M. (2002). Fisher's randomization test and Darwin's data: A footnote to the history of statistics. *Mathematical Biosciences, 180*, 23–28.

Keller, B. (2012). Detecting treatment effects with small samples: The power of some tests under the randomization model. *Psychometrika, 77*, 324–338.

Kempthorne, O. (1952). *The Design and Analysis of Experiments.* New York: Wiley.

Kempthorne, O. (1976). Comment on "On Rereading R. A. Fisher" by L. J. Savage. *Annals of Statistics, 4*, 495–497.

Kempthorne, O. (1979). Sampling inference, experimental inference and observation inference. *Sankhyā: The Indian Journal of Statistics, Series B, 40*, 115–145.

Kempthorne, O., & Doerfler, T. E. (1969). The behavior of some significance tests under experimental randomization. *Biometrika, 56*, 231–248.

Koch, G. (Ed.) (1982). *Exchangeability in Probability and Statistics.* Amsterdam, The Netherlands: North Holland.

Kratochwill, T. R., & Levin, J. R. (2010). Enhancing the scientific credibility of single-case intervention research: Randomization to the rescue. *Psychological Methods, 15*, 124–144.

Krauth, J. (1988). *Distribution-Free Statistics: An Application-Oriented Approach.* Amsterdam, The Netherlands: Elsevier.

Lehmann, E. L., & D'Abrera, H. J. M. (2006). *Nonparametrics: Statistical Methods Based on Ranks* (rev. ed.). New York: Springer.

Lehmann, E. L., & Romano, J. P. (2005). *Testing Statistical Hypotheses* (3rd ed.). New York: Springer.

Lehmann, E. L. (2011). *Fisher, Neyman, and the Creation of Classical Statistics.* New York: Springer.

Lindley, D. V., & Novick, M. R. (1981). The role of exchangeability in inference. *Annals of Statistics, 9*, 45–58.

Manly, B. F. J. (2007). *Randomization, Bootstrap and Monte Carlo Methods in Biology* (3rd ed.). Boca Raton, FL: Chapman & Hall/CRC.

Mardia, K. V. (1990). Obituary: Professor B. L. Welch. *Journal of the Royal Statistical Society, Series A, 153*, 253–254.

Mehta, C. R., & Patel, N. R. (1983). A network algorithm for performing Fisher's exact test in r × c contingency tables. *Journal of the American Statistical Association, 78*, 427–434.

Mehta, C. R., Patel, N. R., & Senchaudhuri, P. (1988). Importance sampling for estimating exact probabilities in permutational inference. *Journal of the American Statistical Association, 83,* 999–1005.

Mehta, C. R., Patel, N. R., Senchaudhuri, P., & Tsiatis, A. A. (1994). Exact permutational tests for group-sequential clinical trials. *Biometrics, 50,* 1042–1053.

Merton, R. K. (1968). The Matthew effect in science. *Science, 159*(3810), 56–63.

Mewhort, D. J. K. (2005). A comparison of the randomization test with the *F* test when error is skewed. *Behavior Research Methods, 37,* 426–435.

Moir, R. (1998). A Monte Carlo analysis of the Fisher randomization technique: Reviving randomization for experimental economists. *Experimental Economics, 1,* 87–100.

Moses, L. E. (1952). Non-parametric statistics for psychological research. *Psychological Bulletin, 49,* 122–143.

Neyman, J. (1923). On the application of probability theory to agricultural experiments: Essay on principles, section 9. *Roczniki Nauk Rolniczych Tom X,* 1–51 (Annals of Agricultural Sciences) Translated and edited by D. M. Dabrowska and T. P. Speed from the Polish original, Statistical Science, 1990, Vol. 5, No. 4, 465–480.

Neyman, J., with cooperation of K. Iwaskiewicz and St. Kolodziejczyk (1935). Statistical problems in agricultural experimentation (with discussion). *Journal of the Royal Statistical Society, Series B, 2,* 107–180.

Neyman, J. (1950). *First Course in Probability and Statistics.* New York: Henry Holt.

Nichols, T. E., & Holmes, A. P. (2002). Nonparametric permutation tests for functional neuroimaging: A primer with examples. *Human Brain Mapping, 15,* 1–25.

Noreen, E. W. (1989). *Computer-Intensive Methods for Testing Hypotheses: An Introduction.* New York: Wiley.

Oden, A., & Wedel, H. (1975). Arguments for Fisher's permutation test. *The Annals of Statistics, 3,* 518–520.

Onghena, P., & Edgington, E. S. (2005). Customization of pain treatments: Single-case design and analysis. *The Clinical Journal of Pain, 21,* 56–68.

Onghena, P., & May, R. B. (1995). Pitfalls in computing and interpreting randomization test *p* values: A commentary on Chen and Dunlap. *Behavior Research Methods, Instruments, & Computers, 27,* 408–411.

Pesarin, F., & De Martini, D. (2002). On unbiasedness and power of permutation tests. *Metron: International Journal of Statistics, 60,* 3–19.

Pesarin, F., & Salmaso, L. (2010). *Permutation Tests for Complex Data: Theory, Applications and Software.* Chichester, UK: Wiley.

Petrondas, D. A., & Gabriel, K. R. (1983). Multiple comparisons by rerandomization tests. *Journal of the American Statistical Association, 78,* 949–957.

Phipson, B., & Smyth, G. K. (2010). Permutation *p*-values should never be zero: Calculating exact *p*-values when permutations are randomly drawn. *Statistical Applications in Genetics and Molecular Biology, 9*(1) [Online] https://doi .org/10.2202/1544-6115.1585.

Pitman, E. J. G. (1937a). Significance tests which may be applied to samples from any populations. *Journal of the Royal Statistical Society, Series B, 4,* 119–130.

Pitman, E. J. G. (1937b). Significance tests which may be applied to samples from any populations. II: The correlation coefficient. *Journal of the Royal Statistical Society, Series B, 4,* 225–232.

Pitman, E. J. G. (1938). Significance tests which may be applied to samples from any populations. III: The analysis of variance test. *Biometrika, 29,* 322–335.

Pitman, E. J. G. (1982). Reminiscences of a mathematician who strayed into statistics. In J. Gani (Ed.), *The Making of Statisticians* (pp. 112–125). New York: Springer.

Pratt, J. W., & Gibbons, J. D. (1981). *Concepts of Nonparametric Theory.* New York: Springer.

Rao, C. R. (1992). R. A. Fisher: The founder of modern statistics. *Statistical Science, 7,* 34–48.

Rasmussen, J. L. (1989). Computer-intensive correlational analysis: Bootstrap and approximate randomization techniques. *British Journal of Mathematical and Statistical Psychology, 42,* 103–111.

Reid, C. (1982). *Neyman from Life.* New York: Springer.

Romano, J. P. (1989). Bootstrap and randomization tests of some nonparametric hypotheses. *The Annals of Statistics, 17,* 141–159.

Rubin, D. B. (1990). Comment: Neyman (1923) and causal inference in experiments and observational studies. *Statistical Science, 5,* 472–480.

Rubin, D. B. (2005). Causal inference using potential outcomes: Design, modeling, decisions. *Journal of the American Statistical Association, 100,* 322–331.

Salsburg, D. (2001). *The Lady Tasting Tea: How Statistics Revolutionized Science in the Twentieth Century.* New York: W. H. Freeman.

Scheffé, H. (1959). *The Analysis of Variance.* New York: Wiley.

Senchaudhuri, P., Mehta, C. R., & Patel, N. R. (1995). Estimating exact *p* values by the method of control variates or Monte Carlo rescue. *Journal of the American Statistical Association, 90,* 640–648.

Shadish, W. R., Cook, T. D., & Campbell, D. T. (2002). *Experimental and Quasi-Experimental Designs for Generalized Causal Inference.* Boston: Houghton Mifflin.

Siegel, S. (1956). *Nonparametric Statistics for the Behavioral Sciences.* New York: McGraw-Hill.

Siegel, S., & Castellan, N. J., Jr. (1988). *Nonparametric Statistics for the Behavioral Sciences* (2nd ed.). New York: McGraw-Hill.

Simon, J. L., & Bruce, P. (1991). Resampling: A tool for everyday statistical work. *Chance, 4,* 22–32.

Soms, A. P. (1977). An algorithm for the discrete Fisher's permutation test. *Journal of the American Statistical Association, 72,* 662–664.

Speed, T. P. (1990). Introductory remarks on Neyman (1923). *Statistical Science, 5,* 463–464.

Sprent, P. (1994). E. J. G. Pitman, 1897–1993. *Journal of the Royal Statistical Society, Series A, 157,* 153–154.

Stigler, S. M. (1978). Mathematical statistics in the early states. *Annals of Statistics, 6,* 239–265.

Stigler, S. M. (1986). *The History of Statistics: The Measurement of Uncertainty before 1900.* Cambridge, MA: Harvard University Press.

"Student" (1923). On testing varieties of cereals. *Biometrika, 15,* 271–293.

ter Braak, C. J. F. (1992). Permutation versus bootstrap significance tests in multiple regression and ANOVA. In K.-H. Jöckel, Rothe, G., & Sendler, W. (Eds.), *Bootstrapping and Related Resampling Techniques* (pp. 79–86). Berlin, Germany: Springer.

Tukey, J. W. (1993). Tightening the clinical trial. *Controlled Clinical Trials, 14,* 266–285.

Vandenbroucke, J. P. (2004). When are observational studies as credible as randomised trials? *Lancet, 363,* 1728–1731.

Weerahandi, S. (1995). *Exact Statistical Methods for Data Analysis.* New York: Springer.

Weerahandi, S. (2004). *Generalized Inference in Repeated Measures: Exact Methods in MANOVA and Mixed Models.* New York: Wiley.

Welch, B. L. (1937). On the z-test in randomized blocks and Latin squares. *Biometrika, 29,* 21–52.

Welch, B. L. (1947). The generalization of "Student's" problem when several different population variances are involved. *Biometrika, 34,* 28–35.

Westfall, P. H., & Young, S. S. (1993). *Resampling-Based Multiple Testing: Examples and Methods for p-Value Adjustment.* New York: Wiley.

Wilks, S. S. (1962). *Mathematical Statistics.* New York: Wiley.

Williams, E. J. (1994). Edwin James George Pitman 1897–1993. *Historical Records of Australian Science, 10,* 2.

Willmes, K. (1987). *Beiträge zu Theorie und Anwendung von Permutationstests in der uni- und multivariaten Datenanalyse.* Unpublished doctoral dissertation, Mathematische-Naturwissenschaftliche Fakultät, Universität Trier.

Winkler, A. M., Ridgway, G. R., Webster, M. A., Smith, S. M., & Nichols, T. E. (2014). Permutation inference for the general linear model. *Neuroimage, 92,* 381–397.

Yates, F. (1984). Tests of significance for 2 × 2 contingency tables (with discussion). *Journal of the Royal Statistical Society, Series A, 147,* 426–463.

Zieffler, A. S., Harring, J. R., & Long, J. D. (2011). *Comparing Groups: Randomization and Bootstrap Methods Using R.* Hoboken, NJ: Wiley.

15

Flexible Minimization: Synergistic Solution for Selection Bias

Donald R. Taves

CONTENTS

15.1 Introduction 229
15.2 Replacing Blocks with Minimization 231
15.3 The Response of Minimization to Selection Bias 232
15.4 Flexible Minimization 234
15.5 Prospective Search for Selection Bias 235
15.6 More Efficient Algorithms 235
15.7 Discussion and Conclusions 236
References 238

15.1 Introduction

In clinical trials, it is theoretically possible to catch and correct selection bias retrospectively in the analysis stage using the Berger–Exner test (Berger and Exner 1999). However, there are few* reports of its having been used in the 18 years since its publication. Therefore, we have only subjective opinions about whether selection bias is an important problem resulting in a wide various of advice from statisticians on the proper design of clinical trials. They range from using only randomization and correcting for imbalances in the analyses (Peto et al. 1976) to reducing it to perhaps less than 10%[†] of the time with minimization.[‡] The reason for the lack of use of the Berger–Exner test is unknown. It could be because of doubts that selection bias is really a problem even when blinding is incomplete or that all of our attention is on the hoped-for publishable positive findings. More likely, collecting appropriate data after the trial is finished may be too difficult and collecting relevant

---

* As of April 2017, only 19 clinical trials from 1999 through 2013 cite the Berger–Exner test and none since.
† An estimate unsupported by any published data.
‡ Capitalization within a sentence refers to the specific algorithm or type of minimization used in Taves (1974b); lowercase refers to the generalized algorithm of Pocock and Simon (1975).

data prospectively is probably too costly if done solely to detect selection bias (assignments for personal advantage, i.e., cheating). It would need to be done for additional reasons.

The mutual interest we have with a cheater is identification of selection covariates. A selection covariate is a combination of variates that must have a sufficient correlation with the outcome to affect it significantly (Berger 2005). We all want to be able to better predict the outcome of treatment, the cheater so he can ensure that the results of the trial support his interests, ours to better serve clinicians in their application of the results of the trial to their patients and to prevent the cheater from succeeding. However, for a cheater to succeed, we have to stay ignorant of the selection covariate that he has discovered. Otherwise, we will use the selection covariate to set up randomized blocks and automatically correct for any imbalance in the primary analysis of the data. If we discover it in the analysis phase, we would use it in the secondary analyses and at least be aware that the data have probably been compromised (Berger 2005).

A strategy that is more likely to succeed in giving proper attention to selection bias is to prospectively collect as much data as possible on potential covariates and to minimize the differences between treatment groups. As the number of patients with the same combination of variates start to become significant they should be treated in the same manner. If our selection of variates or their combinations overlap the selection covariate, we will then start making it difficult for would-be cheaters, more likely finding their selection covariate and, if there is no cheater, then at least improving our ability to predict the outcome and to advance personalized medicine. Using that strategy, however, requires us to reexamine the currently accepted dogma that if you balance a potential variate, you must also use it in your primary statistical analysis (Lin et al. 2015).

Lin et al. (2015) falsely claim universal support for the mantra handed down from Fisher (Senn 2004), namely, that the analysis of the clinical trial must follow the design. Gosset (1937) acknowledged Fisher's concern that balancing without taking it into account would miss some small real differences, but he also noted that it would make up for this by missing fewer large real differences. Subsequent elaboration of Gosset's claim (Pierson 1938; Taves 2001) and references to them have generally been ignored except by Taves (Stigsby and Taves 2010; Taves 2012a,b, 2014), who uses them to advise two routine types of analyses following all clinical trials, particularly after minimization has been used for large numbers of variates of unknown importance. What follows is based on the assumption that it is better to compromise on the "precision" of the p values and use opposite limiting signs for the primary and secondary analyses and the expectation that catching more large real differences is better than catching small real differences that would require repetition with larger samples.

The statistical compromise needed to retain a useful statistical probability and to use large numbers of possible covariates is to declare at the design stage only the few known covariates and use them in the primary statistical analysis of the difference in the mean responses, reporting this p value or the confidence limits as less than the calculated amount. There are two problems with this suggestion. One is how to make it clear that they were chosen at the time of the design. The suggested solution has been to give those variates more weight than the others in the minimization algorithm (Taves 2001). This solution is not ideal because it reduces the number of variates that can be balanced at the same time. The better choice is to retain the formation of unique subgroups using the known important covariates and replacing randomization that is restricted by blocking within each subgroup with minimization. The important covariates are then used exclusively in the primary statistical analysis while all other variates are used in secondary analyses, the process that is referred to as ransacking. This compromise allows us to use as many variates as we can in our search for a selection covariate that a cheater must have to succeed (Berger 2005). General acceptance of this change could have the immediate benefit of making clinical trial data more accessible for creating large data sets (Taves 2014).

15.2 Replacing Blocks with Minimization

The current dominant clinical trial design is to create distinct subgroups of patients and use a separate restricted randomization for each subgroup. The blocking restricts the randomization as the block fills, making it increasingly predictable, the final assignment being certain. Other chapters in this book demonstrate ways to reduce the frequency of predictable assignments. However, since four covariates define at least 16 subgroups, this is generally the maximum number of variates that can be used. There are two limitations that can make even that number impractical. If each subgroup ends with an odd number of patients and the difference assigned to treatment and control happens to be all in the same direction, then chronological bias, the main reason for using the design, becomes a problem. The other limitation is that it requires enough patients to adequately populate all 16 subgroups in order to be efficient and retain enough degrees of freedom. There is a loss of a degree of freedom for each of the covariates in a final analysis of covariance.

Minimization is naturally restricted randomization in the sense that the order that patients become available is unpredictable and hence a form of randomization. It also inherently restricts imbalances in the total number of patients to treatment and control groups (Taves 1974b). We can therefore substitute minimization for the conventional restricted randomization. Using 16 separate minimization algorithms could also result in chronological bias but this is easily rectified. It requires only the addition to each of the 16 algorithms a weighted difference in the total number of patients assigned to treatment and control.

There are three advantages gained from replacing restricted randomization with minimization. The up to four of the known important covariates that would have been heavily weighted are not included, freeing minimization to pay more attention to others. Also, it is easier to maintain a distinction between primary and secondary variates and their analyses at the end of the trial, leaving us freer to change algorithms and variates at any time. Another potential advantage of minimization when someone is cheating is shown in Section 15.3.

15.3 The Response of Minimization to Selection Bias

Selection bias requires a cheater to have a selected patient wait until the next assignment has a high probability of being in the desired group. If the disease being treated is such that delaying treatment is not acceptable, then a cheater has no choice and cannot cheat. If some patients can safely wait, then most patients should be able to wait a small part of the time that some are willing to wait. In the following simulations, that part is set at one fifth of the time. For diseases where more than four times the average waiting time is feasible for a few patients, it is assumed that most patients can wait for the average delay between patients. Having patients wait long enough for a second patient to be assigned at the same time is highly desirable because it makes minimization more efficient (Taves 1974b). Thus, in these simulations, if the cheater has no opportunity to assign the held patient within four average delays between patients, the patient is lost to the simulation.

Obviously, a cheater restricting his selected patients for assignment to one treatment group will start causing an imbalance in the total number of that type of patient between the groups. Minimization will respond to the imbalance by assigning identical patients of non-cheaters to the control group.

The simulations reported below had two investigators, each receiving 1000 patients with equal numbers, 83 or 84, of each of the 12 types formed from having three variates with a total of 7 subcategories. Twenty randomly determined sets of these investigator–patient pairs including the results of the tie-breaking random number generator* were used in simulations with 0, 1, or 2 cheaters. Each cheater had up to five turns to place his select type of patient in the treatment group. With one cheater, minimization assigned the 83 non-cheater's patients of that type to the control group 48.4 ± 3.6 s.d. and 34.6 ± 3.6 s.d. to the treatment group, the t value of the difference = 12. This partially offset the imbalance that the cheater was able to induce with the X number of patients he was able to selectively place.

---

* The tie breaking sequence makes it possible to exactly reproduce the results.

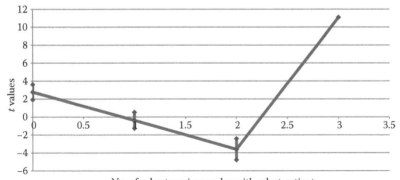

No. of subcategories overlap with select patient
means of t values $\pm$ s.d., n = 2, 5, 4, 1 of grouped t values
12 imbalance differences of 1 vs 0 cheaters, 2 investigators

FIGURE 15.1
Minimization's response to selection bias.

Figure 15.1 shows that it is not only the identical patient types that are affected. The data are the t values* obtained from the simulation with no cheater versus one cheater using the differences in numbers assigned to treatment minus the controls for each of the 12 patient types formed by having 7 subcategories from 3 variates. Clustering was apparent by scanning the results based on whether there were 2, 1, or 0 subcategories in common with the selected type. Therefore, those in each of the three degrees of overlap in subcategories were treated as a group with a mean and standard deviation of their individual t values with an n of 4, 5, and 2, respectively. It shows that those with an overlap of two were also placed more often in the group opposite to the selected patient. Presumably, the two patient types with no overlap in subcategories are a secondary response compensating for response to the ones that had two subcategories in common with the selected patient. The data for both investigators cheating versus none cheating showed the same pattern but more exaggerated. Of those 20 runs, 18 showed the expected signs for the six subcategories; the exceptions had 5 of the 6. The corresponding numbers for one cheater were 10 (6) and 7 (5), the rest were 4s. As of June 17, 2017, the algorithm used for obtaining these data is available at jaytaves .github.io/Minimization using a Chrome web browser. The 20 sets of "Import settings" that are required to exactly reproduce the data will be placed in a more public venue but until then are available from dtaves@pickatime.com.

The implication for the control of selection bias is twofold. Seeing a similar pattern of an unexpectedly large number of variates being unbalanced in a similar relationship to each other would suggest that selection bias was occurring. It could also decrease the cheater's ability to unbalance

---

* t = $(\text{mean}_0 - \text{mean}_1)/\text{sqrt}((\text{Stder}_0{}^*\text{Stder}_0) + (\text{Stder}_1{}^*\text{Stder}_1))$.

his selection covariate if the variates used in minimization had some correlation with elements within the selection covariate, as we hope to have accomplished. Correlated variates tend to act together as though they are more heavily weighted (Taves 1974b). This could make the cheater's task more difficult even if we are unable to identify his selection covariate specifically.

15.4 Flexible Minimization

Flexible minimization is an extension of the generalized algorithm defined by the equations of Pocock and Simon (1975) for finding the best overall balance, that is, G = sum of the d differences for i number of variates subdivided into j subcategories and assigned to k number of treatment groups. Flexible minimization also takes into account the balance of the ij combinations and selects a limited number of the most important of them. The estimates of importance for the variates or combinations will shift after each assignment and add another layer of complexity that the cheater must follow exactly in order to know what the next assignment will be. A sigmoid or uniform random number generator for each assignment can decide the number within specified limits if it is desired to have some degree of randomization in every assignment. This level of complexity would leave the cheater with only probabilistic estimates of what the next assignment will be. He would be left with the fact that all assignment strategies that restrict the difference in overall numbers in the treatment groups become more predictable as the existing difference becomes greater (Barbáchano et al. 2008). The maximum imbalance will depend on the algorithm and the number of patient assignments. For 1000 patients, a maximum of 4 was observed with Minimization and 8 with Rank-Minimization (Stigsby and Taves 2010).

This means that the cheater would probably be faced by differences in the number of patients ranging from at least 0 to 4 to 0 to 8. The overall probability of correctly guessing the next assignments would be 0.683 ± 0.048 and 0.636 ± 0.059, respectively, the latter being less predictable than with randomized blocks of 5 (Stigsby and Taves 2010). He then has the dilemma of risking submitting a sufficient number of his select patients with a high risk of their being assigned to the non-selected treatment, or waiting for the rare instance with a high probability that the assignments will be to the desired treatment. In other words, the higher the probability of correctly guessing the next assignment, the longer the investigator would have to be willing to wait to get the assignment desired. If the patients can wait, we can foil the cheater's efforts by having all patients, or just the ones suspected of being used for selection bias, wait for one or more additional patients to be presented before they are assigned.

15.5 Prospective Search for Selection Bias

The cheater's chances of causing selective placement can be further decreased by monitoring the high probability assignments. If a particular type of patient or a particular investigator is represented more than the overall average, the investigator can be warned by the computer that assignment of that patient is suggestive of selection bias. He can also be informed that the suspicion would probably be removed if the patient could wait until another patient were available for assignment. If a delay was unacceptable, he could be given the option of calling for a human monitor to okay immediate assignment.

As the clinical trial progresses and output data start to become available, the computer can be programmed to look for an association of the outcome data with the highly predictable assignments, that is, the Berger–Exner test. Further evidence of selection bias can be met with direct intervention. A suspected patient's chart or central laboratory records can be reviewed for data on variates other than those supplied for minimization. These can give hints on a selection covariate. Lack of cooperation in disproving the suggestive evidence would be damaging information about an investigator and put him in the position of having to prove he was not engaged in selective placement. This would mean a shift from our current assumption that it is our job to show convincing evidence that an investigator is cheating. Retrospectively, an investigator has very limited ways of demonstrating his innocence and can claim that randomization is bound to deliver some false suggestive imbalances. These options in addition to the systematic search for evidence of selection bias with Flexible minimization and its large number of variates being monitored for imbalance will make it unlikely that a would-be cheater would take the risk of getting caught.

15.6 More Efficient Algorithms

The Minimization algorithm originally described and probably still the most employed in the 2% to 4% of clinical trials (Higham et al. 2015; Taves 2010) that use them today is inefficient to use in Flexible minimization. There are algorithms that pay much more attention to the variates that are the most unbalanced, such as squaring the differences rather than summing the absolute differences. The absolute differences can be multiplied by the squared t value of the original data. Both modifications reduce the number of severely imbalanced variates (difference in means showing t values greater than 2) relative to the original algorithm. Compared to their controls, the numbers were reduced by a factor of 0.19 when summing the absolute differences, by 0.08 when squaring the differences, and by 0.02 squaring the t value of the original data (Taves 1974a). These algorithms fit the generalized equation for minimization described by Pocock and Simon (1975).

This family of algorithms is considered Partial minimization because they use only the subcategories that the incoming patient has to make the assignment decision. Complete minimization uses optimization algorithms that use all the data of all the patients for each assignment. The necessary computer calculations for Complete minimization cannot be done until there are as many patients who have been entered as there are variates to be balanced. These constraints limit the number of variates that can be used and delay when they can start balancing the groups relative to partial minimization algorithms (Taves 2010). However, they might prove useful under some circumstances or after Flexible minimization has already assigned sufficient patients.

The increase in the number of variates gained by the above changes in partial minimization will probably be sufficient for most of the clinical trials run today. However, when functional genomic data are available, the known predictive combinations of weak covariates would probably be used rather than the individual variates. These are of course what the successful cheater has to know and we wish we did.

15.7 Discussion and Conclusions

The belief that selection bias can only be controlled with randomization or complete blinding is no longer tenable. Flexible Minimization with prospective searching for evidence of selection bias is an alternative where both goals can be advanced at the same time. In theory, it can be controlled better than it has been in several ways. However, we are not likely to be able to prove it because we lack data even proving that selection bias in the assignment process exists, much less quantifying it. Even if we do not find a selection covariate, we will have better data immediately in hand to retrospectively search for it, or for evidence of selection bias. This should in turn improve the reporting of the efforts to prevent and detect selection bias, something that the authors of this book think should be done. Making the suggested changes will probably stop anyone from trying it because of less payoff with increased risk of getting caught, which is reason enough for doing it. But the prospects of improving our search for more powerful predictive covariates or combinations should also be enhanced, which is reason enough for further exploration of using Flexible minimization, even if selection bias is no longer an issue.

The immediate challenge for most researchers will be the belief that the statistical analysis must follow the design. They, following Fisher's thinking (Ziliak and McCloskey 2011), are correct that we will be giving up the precise p values and fail to detect some small real effects. However, in exchange for that, as pointed out by "Student," i.e., Gosset (1937), we gain by missing fewer large real effects, that is, steeper power curves (Pierson 1938; Taves 2001). Steeper power curves mean better repeatability of clinical trials. Those benefits are germane because of the recent concern about the lack of repeatability of clinical trials and their high cost relative to their benefits (Devereaux and Salim 2013; Lauer 2015).

Even recent articles primarily concerning the statistical analysis after minimization (Lin et al. 2015; Shao and Ur 2013) fail to consider the proposal of using *greater-than* signs for minor or questionable variates that show some retrospective association after minimization. Flexible minimization goes a step further by changing the variates and their weighting during the course of the trial. It would not be expected to give precise p values in a theoretically infinite repetition. However, obtaining precise p values in the secondary analyses is difficult if not impossible, regardless. The straightforward way to do it is the Bonferroni correction that requires knowing how many hypotheses pass through our minds, which is not easy, to say nothing of reading someone else's mind before hitting on a combination of variates that appear to be associated (Wikipedia: Bonferroni correction and post hoc analysis, February 24, 2017). We would be squandering our resources to rely on them as the sole basis for starting another clinical trial. However, the p values could be as low as those calculated, and finding them is likely to be valuable in getting us to consider other reasons for why the possible significant combination might have value.

There are other advantages that should be kept in mind. Currently, most investigators are unwilling to share the individual data from their completed clinical trials (Berlin et al. 2014; Ferguson and Hebert 2014). Their concern is that "parasites" (McNutt 2016) will ransack the data in order to discredit the conclusions of the original investigators and thereby weaken the value of their reports (Largo and Drazen 2016). These concerns could be lessened if it were generally accepted that any analyses by others beyond a check on the primary analysis would be considered post hoc investigations, requiring any probability statements they assign about the null hypotheses reported with limiting *greater-than* signs. This should decrease the fear that p values from post hoc analyses will be overvalued. Hopefully, this would result in a greater willingness to share individual data from completed clinical trials to form large databases for our enlightenment.

We should also encourage investigators to think about other factors that might make a difference in predicting the outcome (Feinstein 1972; Horwitz et al. 1998). Letting them add variates to the protocol at any time so they can ransack the data after the trial is done to see if they can see any support for their ideas will keep them involved and make use of their inquisitiveness and experience.

Clinical trials can be valuable as a source of information on the predictors of the outcome of disease, regardless of whether the new treatment being tested has any effect. Having that as one of the purposes of a clinical trial can ease the feeling of failure when the treatment does not show the expected effect. This secondary purpose, of course, fits well with the suggestions of improving our prevention and detection of selection bias. Finding new or better combinations for predicting the outcome of treatment maximizes our ability to detect a selection covariate that a cheater might exploit.

The most likely adverse effects of adopting Flexible minimization and the limiting signs is forgetting that the limiting signs are not absolutes for judging individual studies. They will still only be statistical statements that are the best we can do with the data at hand and that further study may change our minds. Ideally, we should become even more systematic in our control of errors, like they do in laboratory method research, so we are less dependent on statistical calculations and statements in making sound conclusions (Ziliak and McCloskey 2011). These possible benefits together with a synergistic effect for improving our search for predictive combinations of variates, reducing the risk of selection bias, and improving the repeatability of studies make further study of Flexible minimization highly desirable.

We do need more simulations to better understand the potential and limits of Flexible minimization. Specifically, how many more variates can be profitably used in Flexible minimization? Judging the results will require some agreed-upon goal(s) such as using the ability of minimization in the original simulations to decrease the imbalance of 15 diverse variates with a fivefold reduction of simulations with t values of 2 and higher. A less complex indicator is the frequency with which Flexible minimization uses a random number generator to break tie scores. Those assignments do not improve the balance and therefore are wasted from the point of view of wanting to obtain the best possible balance of the treatment groups. Using the random number generator to select another set of the most important variates for Flexible minimization could be desirable to reduce the frequency of use of a random number generator to break tie scores.

Hopefully, pursuit of alternative unbiased methods of impartial assignments of patients to treatment groups will lead to a consensus among statisticians so they can speak with one voice that will provide better guidance in improving clinical trials.

References

Barbáchano Y, Coad DS, Robinson DR. Predictability of designs which adjust for imbalances in prognostic factors. *J Stat Plan Inference* 2008; 138: 756–767.

Berger VW. *Selection Bias and Covariate Imbalances in Randomized Clinical Trials.* John Wiley & Sons Ltd. The Atrium, Southern Gate, Chichester, West Sussex PO19 8SQ, England, 2005.

Berger VW and Exner DV. Detecting of selection bias in randomized clinical trials. *Control Clin Trials* 1999; 20: 319–327.

Berlin JA, Morris S, Rockhold FA et al. Bumps and bridges on the road to responsible sharing of clinical trial data. *Clin Trials* 2014; 11: 7–12.

Devereaux PJ and Salim Y. Editorial: When it comes to trials, do we get what we pay for? *N Engl J Med* 2013; 369: 1962–1963.

Feinstein AR. Clinical biostatistics. XXIV. The purposes of prognostic stratification. *Clin Pharmacol Ther* 1972; 13: 285–297.

Ferguson DA and Hebert PC. Commentary on Berlin et al. *Clin Trials* 2014; 11: 13–14.

Higham R, Tharmanathan P, Birks Y. Use and reporting of restricted randomization: A review. *J Eval Clin Practice* 2015; 21(6): 1205–1211.

Horwitz RI, Singer BH, Makuch RW et al. Response to Senn S and Harrell F. Subgroups and groping for significance. *J Clin Epidemiol* 1998; 51(12): 1367–1368.

Largo DL and Drazen JM. Data sharing. *New Engl J Med* 2016; 374(20): 276–277.

Lauer MS. *"A Future for Clinical Trials? An NIH Perspective"* The Founder's lecture SCT 36th Annual Meeting (2015) online at http://www.sctweb.org/public /meetings/2015/, July 5, 2017.

Lin Y, Zhu M, Su Z. The pursuit of balance: An overview of covariate-adaptive randomization techniques in clinical trials. *Contemp Clin Trials* 2015; 45(Supplement PA): 21–25.

McNutt M. #IAmAResearchParasite. *Science* 2016; 351(6277): 1005.

Peto R, Pike MC, Armitage P et al. Design and analysis of randomized clinical trials requiring prolonged observation of each patient. I. Introduction and design. *Br J Cancer* 1976; 34: 585–612.

Pierson ES. Some aspects of the problem of randomization*: II. An illustration of 'Student's' inquiry into the effect of 'balancing' in agricultural experiments. *Biometrika* 1938; 30: 159–179.

Pocock SJ and Simon R. Sequential treatment assignment with balancing for prognostic factors in the controlled clinical trial. *Biometrics* 1975; 31(1): 103–115.

Senn S. Added values: Controversies concerning randomization and additivity in clinical trials. *Statist Med* 2004; 23: 3729–3753.

Shao J and Ur X. Validity of test under covariate-adaptive biased coin randomization and generalized linear models. *Biometrics* 2013; 69(4): 960–969.

Stigsby B and Taves DR. Rank-minimization for balanced assignment of subjects in clinical trials. *Contemp Clin Trials* 2010; 31(2): 147–150.

'Student' (Gosset). Comparison between balance and random arrangement of field plots. *Biometrika* 1937; 29: 363–379.

Taves DR. Alternative methods of minimization: Linear, exponential and statistical. Not published, 1974a.

Taves DR. Minimization: A new method of assigning patients to treatment and control groups. *Clin Pharmacol Ther* 1974b; 15(5): 443–453.

Taves DR. Optimum biased-coin designs for sequential treatment allocation with covariate information by A. C. Atkinson. *Stat Med* 2001 Mar 15; 20(5): 813–818.

Taves DR. Rank-minimization with a two-step analysis should replace randomization in clinical trials. *J Clin Epidemiol* 2012a; 65: 3–6.

Taves DR. Rebutal of anon's critique: In fact randomization is the limiting probability distribution for imbalance with minimization. *J Clin Epidemiol* 2012b; 65: 811.

Taves DR. Making data from clinical trials available. *Clin Trials* 2014; 11(6): 685.

Taves DR. The use of minimization in clinical trials. *Contemp Clin Trials* 2010; 31(2): 180–184.

Ziliak ST and McCloskey DN. *The Cult of Statistical Significance: How the Standard Error Costs Us Jobs, Justice and Lives.* The University of Michigan Press. Ann Arbor, Michigan. 2011.

Index

Page numbers followed by f, t, and n indicate figures, tables, and footnotes respectively.

A

ACE Inhibitor After Anthracycline study, 58
Adaptive Randomization 2017, 11
Additional reference statistics, 196
AESB, *see* Average excess selection bias (AESB)
Affection, 30
"Allocation bias," 13–14
Allocation concealment, 85
    dual threats to, 129–130
Allocation Ratio Preserving (ARP) procedures, 92
Alternative allocation, 2, 3, 4
Alternative hypothesis, of randomization test, 193–195
AMP, *see* Asymptotic maximal procedure (AMP)
Angiotensin-converting enzyme (ACE), 58
Approbation, 30
ARP (Allocation Ratio Preserving) procedures, 92
Assumptions, distributional, 201
Asymptotic maximal procedure (AMP), 77; *see also* Maximal procedure (MP)
    allocation sequences, 77, 77f
    conditional allocation probability for, 67, 69, 69t, 71–72, 72f, 73t–74t
Average excess selection bias (AESB), 106
    calculation, 107, 111–113, 112t, 113f, 114
    under directional strategy, 114–115
        115t, 116t, 117–118, 118t–119t

B

Balanced Urn Design, 54
Barnard, George A., 220

BCDWIT, *see* Biased coin design with imbalance tolerance (BCDWIT)
Behavior(s)
    leading to bias, 31–32
Berger–Exner test, 85, 143, 229, 235
    third-order selection bias detection, 159–169
        background, 159–160
        bias testing, 163–164
        data analysis, 164
        discussion, 165–169
        limitations and strengths of method, 165–167
        methods, 160–164
        recommendation for further research, 169
        results, 164, 165t, 166t, 167t
        study scenarios and parameters, 162–163, 163t
        trial simulation, 161–162
Bias(es), 13; *see also specific types*
    behaviors leading to, 31–32
    classification of risk of, 84
    confounding, 53, 58
    confusion, and inappropriate randomization process, association between, 14–21, 16t–21t
    elimination, 187–188
    expected bias factor, 95–97
    historical perspective, 1–6
        Austin Bradford Hill and strepto-mycin trial in tuberculosis treatment (1948), 3–5
        early twentieth-century discussions, in clinical research, 2–3
        "lessons learned" and permuted block design, 5–6

investigator preferences, 29–32
motivations for, 29–30
overview, 1
restricted randomization to control,
    53–54
    covariate-adjusted, 55–56
selection, synergistic solution for,
    see Selection bias
testing, 145–146, 163–164
Bias-creating behaviors, 31
Biased coin design with imbalance
    tolerance (BCDWIT), 78
conditional allocation probability
    for, 65
Biased Coin Randomization, 53, 54, 57
Big stick design (BSD)
conditional allocation probability
    for, 64
Blackwell-Hodges model, 90, 91, 172
for selection bias in studies with
    unequal allocation, 91–97
elements, 91–94
expected bias factor, 95–97
Block randomization, 53
Blocks
replacing, with minimization,
    231–232
Block urn design (BUD), 77
allocation sequences, 77, 77f
conditional allocation probability
    for, 64
Bonferroni correction, 237
Bootstrapping, framework of, 218
Breast Cancer Erythropoietin Trial,
    the, 58
Brick Tunnel Randomization (BTR), 91
PBR in study with, selection bias,
    116–119, 117f, 118t–119t
British Medical Journal, 2
British War Office, 2
BSD, see Big stick design (BSD)
BTR, see Brick Tunnel Randomization
    (BTR)
BUD, see Block urn design (BUD)

C

Causal inference, 201–202
Ceteris paribus, 12–13, 12n

Chalmers, Iain, 3
Cheater, selection covariates and, 230–235
Chronological bias
    vs. selection bias, 128–129
Churchill, Winston, 125
Clarification, of equating property of
    randomization, 188–189, 191
Clinical equipoise's principle, 12, 12n
Clinical trials
    purposes, 237–238
    randomization in, 186–188
    selection bias, 229–230
clinicaltrials.gov, 53
Cochrane Collaboration, 10, 142
Collective equipoise, 30
Colophon, 23–24
Comparability, lack of, 14, 15, 18t–19t, 21
Comparison groups, in RCTs
    high sample size impact on balance
        of, 15, 20t–21t, 21
    increasing sample size impact on
        balance of, 15, 18t–19t
    unbalance of, 15, 16t–17t
Complete minimization, 236
Compliance
    elements of, 31
Computer-intensive tests,
        randomization tests and, 218
Conditional allocation probability
    restricted randomization evolution
        with MTI procedure, 62–76
    asymptotic maximal procedure
        (AMP), 67, 69, 69t, 71–72, 72f,
        73t–74t
    biased coin design with
        imbalance tolerance
        (BCDWIT), 65
    big stick design (BSD), 64
    block urn design (BUD), 64
    comparison of, 72, 75–76, 75t
    maximal procedure (MP), 65–67,
        66f, 68t
    permuted block design (PBD), 63
Conditional inference, 201–202
Confounding, 14
    bias, 53, 58
Confounding variables
    measurement, 188–189
    symmetrically distributed, 191

Confusion bias, 14
   and inappropriate randomization
      process, association between,
      14–21, 16t–21t
Convergent strategy, 173
Covariate-adjusted response-adaptive
      randomization, 51
Covariate-adjusted restricted
      randomization
   case studies, 57–59
   cautions (considerations) of, 56–57
   to control biases, 55–56
Critical rationalism, 15
Customized test statistics, 200–201

D

Darwin, Charles, 192, 210
Defective randomization, 84–85
Defect number, 84
The Design of Experiments, 9, 192, 210, 211,
      212, 216
Design vector, randomization test, 195
Dictionary of Epidemiology, 10, 11
Directional guessing strategy
   AESB under, 114–115, 115t, 116t,
      117–118, 118t–119t
   exact conditional, justification of,
      97–99
   expected bias factor under, 105–107, 114
   justification of, 100–104, 103f
   with payoff threshold, 113–115,
      114f–116f
   for PBR, 102–103, 103f
Distributional assumptions, 201
Double-blinded clinical trials, 171
Double-blind element, 1
Dual threats, Unrecognized, *see*
      Unrecognized dual threats, to
      internal validity
Duck, Daffy, 126
Dvorak keyboard, 129

E

Early twentieth-century discussions,
      of bias and randomization in
      clinical research, 2–3
Encyclopedia of Statistical Sciences, 217

Equating property of randomization
   clarification, 188–189, 191
   concluding remarks, 190–191
   example, 190
   statistical regularity, 192
Equipoise
   collective, 30
   defined, 12n
Errors
   classical theory of, 211
   Gaussian theory of, 212
   normal law of frequency of, 210
   theory of, 210
Ethics
   clinical, 30
   of clinical trials, randomization and, 6
European Organisation for Research
      and Treatment of Cancer
      Working Group, 58–59
Evaluation, 83–87
   classification of risk of bias, 84
   defect number, 84
   flawed trials, 124–125
   focus, 83–84
   overview, 83
   studies, defects in, 84–85
   suboptimal studies, consequences to
      using, 86–87, 86t
Evidence, concept of, 10, 10n
Evidence-based medicine, 10
Exact conditional guessing strategy,
      justification of, 97–99
Exact tests, randomization tests and, 219
Exchangeability, randomization tests
      and, 219–220
Expected bias factor, 95–97
   under directional strategy, 105–107,
      114

F

False knowledge, propagation of, 126–128
Falsificationism, 15
Features, randomization test, 199–202
   distributional assumptions, 201
   flexibility with respect to design and
      test statistic, 200–201
   frequentist, conditional, and causal
      inference, 201–202

random sampling, 201
validity and power by construction,
    199–200
Federal Drug Administration
    recommends designing studies, 57–58
Fictional indifference curve, for
    unsympathetic investigator,
    35–36, 36f
Fictional Nash equilibrium, with
    unsympathetic investigator,
    36, 37f
Fisher, R. A., 9, 52, 193, 194, 201, 210–212,
    213, 214, 215
Fisher's Exact Test, 202, 216
Flawed trial evaluations, translate into
    flawed trials, 124–125
Flexibility, randomization test, to any
    design, 200–201
Flexible minimization, 234
Frequency probability, *see* Randomization
Frequentist, 201–202

G

Galton, Francis, 192
Galton's analysis of Darwin's data,
    210–211
Gaussian theory, of errors, 212
GoogleScholar, 160

H

Hill, Austin Bradford, 3
    and streptomycin trial in tuberculosis
        treatment (1948), 3–5
Hippocrates, 10
Historical clarification, 210–216
    Fisher, 210–212
    overview, 210
    Pitman and Welch, 212–214
    Stigler's law of eponymy, 214–216
Historical perspectives, 1–6
    Austin Bradford Hill and streptomycin
        trial in tuberculosis treatment
        (1948), 3–5
    early twentieth-century discussions,
        in clinical research, 2–3
    "lessons learned" and permuted
        block design, 5–6

I

Ignorance *vs.* malice, 131–133
Internal validity
    selection bias effects on, 141; *see also*
        Selection bias
    unrecognized dual threats to,
        123–137
        allocation concealment, dual
            threats, 129–130
        betting odds *vs.* biasing allocation
            only with certainty, 130–131
        chronological bias *vs.* selection
            bias, 128–129
        false knowledge, propagation of,
            126–128
        flawed trial evaluations translate
            into flawed trials, 124–125
        malice *vs.* ignorance, 131–133
        and Omnibus solutions, 135–136,
            135t
        overview, 123–124
        unmasking (early *vs.* late), 133–135
Investigator preferences, for bias, 29–32
    behaviors leading to, 31–32
    motivations, 29–30
Isoprofit curves, preferences regarding
        predictability and sample size,
        32–33, 33f
Iso-utility curves, preferences regarding
        predictability and sample size,
        33–35, 34f

J

Jadad score, 124

K

King, Martin Luther, Dr., 133
Knowledge, false, propagation of,
    126–128

L

Law of eponymy, Stigler, 210, 214–216
Lee, Robert E., 130
Likelihood ratio test (simulation study),
    180–183, 181t, 183t

M

Mahalanobis distance, 55
Malice *vs.* ignorance, 131–133
Matched randomization, 55, 56, 57, 58, 59
Maximally tolerated imbalance (MTI)
 procedure, 54, 129, 130,
 134–135, 161
 history, 61
 overview, 39–40, 61
 restricted randomization evolution
  with, 61–79
  asymptotic maximal procedure
   (AMP), 67, 69, 69t, 71–72, 72f,
   73t–74t
  biased coin design with
   imbalance tolerance
   (BCDWIT), 65
  big stick design (BSD), 64
  block urn design (BUD), 64
  comparison of, 72, 75–76, 75t
  conditional allocation probability,
   62–76
  implementation of, 79
  maximal procedure (MP), 65–67,
   66f, 68t
  overview, 61–62
  permuted block design (PBD), 63
  treatment allocation
   predictability, 76–79, 77f–78f
 selection mechanism
  induced by, 43–45
  value for, 42–43
 use of, 41
Maximal procedure (MP), 54, 77;
  see also Asymptotic maximal
  procedure (AMP)
 allocation sequences, 77, 77f
 conditional allocation probability for,
  65–67, 66f, 68t
Medical Research Council, 4
Merton, Robert, 216
Meta-DiSc Version 1.4 statistical
 software, 146
Minimization algorithm, selection bias
 and
 complete, 236
 flexible, 234
 partial, 236

 replacing blocks with, 231–232
 response, 232–234
Minimization randomization, 56, 57, 58
Modern-day randomization, 52; *see also*
 Randomization
Monte Carlo versions, of randomization
 tests, 219
Motivations
 for bias, 29–30
MTI procedure, *see* Maximally tolerated
 imbalance (MTI) procedure
Multiplicative model, 194
Multivariate linear regression model,
 56

N

Nash equilibrium, 35, 35f, 36, 36f
 fictional, with unsympathetic
  investigator, 36, 37f
Negative falsehoods, 13
Neyman, Jerzy, 193–194, 197, 201, 215
Neyman–Pearson approach, 193–194,
 197, 201
Normal law of frequency of error, 210
Notation, randomization test, 192–193
Null hypothesis, randomization test,
 193–195, 197, 202, 203
Numberless, 24

O

Omnibus solutions, unrecognized dual
 threats and, 135–136, 135t
Online Etimology Dictionary, 10n, 11n
Open-label single-center trials, 89

P

Partial minimization, 236
Patients; *see also* Comparison groups
 strategic selection of, 31
Payoff threshold, directional strategy
 with, 113–115, 114f–116f
PBD, *see* Permuted block design (PBD)
PBR, *see* Permuted block randomization
 (PBR)
Pearl Harbor, attack on, 132
Pearson, Karl, 2–3

Peirce, Charles, 52
Permutation tests, 209–220
    defined, 217
    historical clarification, *see* Historical
        clarification
    overview, 209
    randomization tests *vs.*, 217–218
    terminological clarification, *see*
        Terminological clarification
Permuted block design (PBD), 5–6
    allocation sequences, 76–77, 77f
    conditional allocation probability
        for, 63
    treatment allocation predictability,
        76–77
Permuted block randomization (PBR),
        54, 55, 57, 128
    directional strategy for, 102–103, 103f
    selection bias with, 107–113,
        108t–109t, 110f, 112f, 112t
    in study with BTR, selection bias,
        116–119, 117f, 118t–119t
Pitman, Edwin J. G., 212–213, 215
Placebo effect(s), 32
    defined, 32
Popper's theory, 15
Positive falsehoods, 13
Power, by construction, 199–200
Predictability
    of clinical trial design, preferences
        regarding, 32–36, 33f–37f
        fictional indifference curve, 35–36,
            36f
        fictional Nash equilibrium with
            unsympathetic investigator,
            36, 37f
        isoprofit curves, 32–33, 33f
        iso-utility curves, 33–35, 34f
        Nash equilibrium, 35, 35f, 36, 36f
        response functions, 35, 35f
    treatment allocation, 76–79, 77f–78f
Principles of Medical Statistics, 3
Propagation, of false knowledge, 126–128
Properties
    bias elimination, 187–188
    exchangeability, 220
    randomization, 187–188
Prospective search, for selection bias,
    235

PubMed, 160
Purposes, of randomization, 192
p value, randomization test, 195–197,
    201–202

Q

QWERTY keyboard, 128, 129

R

Random, defined, 11
Random Allocation Software 2017, 12
Randomization, 185–203
    appropriate method selection, 83
    ceteris paribus, 12–13, 12n
    clinical equipoise's principle, 12, 12n
    in clinical trials, 186–188
    colophon, 23–24
    defined, 10–11, 186
    described, 11–12
    equating property
        clarification, 188–189, 191
        concluding remarks, 190–191
        example of, 190
        statistical regularity, 192
    and ethics of clinical trials, 6
    example, 189–190
    experimental condition, 52
    historical perspective, 1–6
        Austin Bradford Hill and strepto-
            mycin trial in tuberculosis
            treatment (1948), 3–5
        early twentieth-century discussions,
            in clinical research, 2–3
        "lessons learned" and permuted
            block design, 5–6
    link between randomization test
        and, 202–203
    modern-day, 52
    overview, 1, 185–186
    phases, 11
    physical act of, 187
    principle, 12
    process
        inappropriate, confusion bias and,
            14–21, 16t–21t
        overview, 11

reporting of, 21–23
selection, 13–14
property of, 187–188
in psychophysical investigations, 185
purpose of, 83
reasons, 187–188
relevance, 12–13
restricted, 51–59; *see also* Restricted randomization
case studies, 57–59
cautions (considerations) of covariate-adjusted approaches, 56–57
to control biases, 53–54
covariate-adjusted, to control biases, 55–56
criticism, 52
evolution of, with MTI procedure, 61–79
historical context, 52–53
overview, 51
schemes, 53; *see also specific entries*
unrestricted, 39–47
facts regarding, 45–47
MTI procedure induced by selection mechanism, 43–45
MTI value for selection mechanism, 42–43
overview, 39–40
theory and practice, 40–42
Randomization.com 2017, 12
Randomization tests, 191–202
computer-intensive tests and, 218
defined, 217
exact tests and, 219
example, 197–198
exchangeability and, 219–220
historical clarification, *see* Historical clarification
link between randomization and, 202–203
main features, 199–202
distributional assumptions, 201
flexibility with respect to design and test statistic, 200–201
frequentist, conditional, and causal inference, 201–202
random sampling, 201

validity and power by construction, 199–200
Monte Carlo versions of, 219
notation, 192–193
null hypothesis and alternative hypothesis, 193–195
overview, 209
permutation tests *vs.*, 217–218
purposes, 192
p value, 195–197
statistic, 195
terminological clarification, *see* Terminological clarification
Randomized controlled trials (RCTs), 1, 160; *see also specific trials*
allocation bias, 13–14
comparison groups in
high sample size impact on balance of, 15, 20t–21t, 21
increasing sample size impact on balance of, 15, 18t–19t
unbalance of, 15, 16t–17t
ethical aspects, 1, 6
flawed trial evaluations translate into flawed trials, 124–125
history, 9–10
methodological aspects, 1
permuted block design, 5–6
philosophy of, 10
RPS-based testing for second-order selection bias effect in, 141–154
bias testing, 145–146
data analysis and summary measures, 146, 147t
discussion, 149–154, 151f–153f
four-step RCT simulation with R code "rps.gen," 156
materials and methods, 143–146
overview, 141–143
recommendation for further research, 154
results, 146–149, 148f, 149t
study method limitations, 149–150
test results (raw data), 157–158
test sensitivity and specificity, 146–147
trial parameters and test accuracy, 149, 149t

trial scenarios, 144–145
trial simulation and parameters, 143–144
selection bias in, adjustment and detection, 171–183
likelihood ratio test performance (simulation study), 180–183, 181t, 183t
model, 172–180
overview, 171–172
Random.org 2017, 12
Random sampling, 201, 210–212, 220
Random variable, 174
Ransacking, defined, 231
Rao, C. Radhakrishna, 52
RCTs, *see* Randomized controlled trials (RCTs)
Reference distribution, 196, 198
Reference set
cardinality of, 193
defined, 193
Regression analysis, 160
Reid, Constance, 215
Reid, D. D., 4
Replacing blocks, with minimization, 231–232
Reports/reporting
inappropriate, 23
randomization process, 21–23
Resampling, framework of, 218
Research
early twentieth-century discussions of bias and randomization in, 2–3
Research Randomizer 2017, 12
Response
functions, preferences regarding predictability and sample size, 35, 35f
of minimization to selection bias, 232–234
vector, randomization test, 195–196
Restricted randomization, 51–59; *see also* Randomization
case studies, 57–59
cautions (considerations) of covariate-adjusted approaches, 56–57
to control biases, 53–54

covariate-adjusted, to control biases, 55–56
criticism, 52
evolution of, with MTI procedure, 61–79
asymptotic maximal procedure (AMP), 67, 69, 69t, 71–72, 72f, 73t–74t
biased coin design with imbalance tolerance (BCDWIT), 65
big stick design (BSD), 64
block urn design (BUD), 64
comparison of, 72, 75–76, 75t
conditional allocation probability, 62–76
implementation of, 79
maximal procedure (MP), 65–67, 66f, 68t
overview, 61–62
permuted block design (PBD), 63
treatment allocation predictability, 76–79, 77f–78f
historical context, 52–53
as Markov process, 62
overview, 51
Reverse propensity score (RPS), 143, 160
testing for second-order selection bias effect in RCTs, 141–154
bias testing, 145–146
data analysis and summary measures, 146, 147t
discussion, 149–154, 151f–153f
four-step RCT simulation with R code "rps.gen," 156
materials and methods, 143–146
overview, 141–143
recommendation for further research, 154
results, 146–149, 148f, 149t
study method limitations, 149–150
test results (raw data), 157–158
test sensitivity and specificity, 146–147
trial parameters and test accuracy, 149, 149t
trial scenarios, 144–145
trial simulation and parameters, 143–144

Risk of bias (ROB), 84
  assessment tool, 142
ROMPA study, 23
Rose, Ernestine L., 126
RPS, *see* Reverse propensity score
    (RPS)

S

Sample size
  balance of comparison groups in
    RCTs
    high size, 15, 20t–21t, 21
    increasing size and, 15, 18t–19t
  and predictability, preferences
    regarding, 32–36, 33f–37f
    fictional indifference curve, 35–36,
      36f
    fictional Nash equilibrium with
      unsympathetic investigator,
      36, 37f
    isoprofit curves, 32–33, 33f
    iso-utility curves, 33–35, 34f
    Nash equilibrium, 35, 35f, 36, 36f
    response functions, 35, 35f
Sampling, random, 201, 210–212, 220
Scott, R. A., 212
Sealed Envelope 2017, 12
Second-order selection bias
  effect in RCTs, RPS-based testing,
    141–154
    bias testing, 145–146
    data analysis and summary
      measures, 146, 147t
    discussion, 149–154, 151f–153f
    four-step RCT simulation with R
      code "rps.gen," 156
    materials and methods, 143–146
    overview, 141–143
    recommendation for further
      research, 154
    results, 146–149, 148f, 149t
    study method limitations,
      149–150
    test results (raw data), 157–158
    test sensitivity and specificity,
      146–147
    trial parameters and test accuracy,
      149, 149t

    trial scenarios, 144–145
    trial simulation and parameters,
      143–144
Selection bias, 14, 39, 40
  effects on internal validity; *see also*
    Selection bias
  in RCTs, adjustment and detection,
    171–183
    likelihood ratio test performance
      (simulation study), 180–183,
      181t, 183t
    model, 172–180
    overview, 171–172
  second-order, *see* Second-order
    selection bias
  in studies with unequal allocation,
    89–120
    Blackwell-Hodges model for, 91–97;
      see also Blackwell-Hodges model
    directional guessing strategy,
      97–107
    directional strategy with payoff
      threshold, 113–115, 115t–116t
    overview, 89–91
    PBR, 107–113, 108t–109t, 110f, 112f,
      112t
    in PBR in study with BTR, 19t,
      116–119, 117f, 118t
    third-order, *see* Third-order selection
      bias
    vs. chronological bias, 128–129
Selection bias, synergistic solution for,
    229–238
  discussion, 236–238
  efficient algorithms, 235–236
  flexible minimization, 234
  overview, 229–231
  prospective search for, 235
  replacing blocks with minimization,
    231–232
  response of minimization to, 232–234
Selection covariates, identification of,
    230, 233–234
Selection effect, 173
Sequence generation, 85
*A Short Treatise on Anti-Typhoid
    Inoculation*, 2
Simplified Acute Physiology Score
    (SAPS) III score, 23

Smith, Adam, 30
Socrates, 10
SPIRIT (Standard Protocol Items:
        Recommendations for
        Interventional Trials)
        statement, 21
SROC curves, *see* Symmetric
        summary receiver operating
        characteristic (SROC) curves
Standard Protocol Items:
        Recommendations for
        Interventional Trials (SPIRIT)
        statement, 21
Stanford Encyclopedia of Philosophy, 10, 11
"Statistical Methods for Research
        Workers," 9
Statistical regularity, 192
Statistical significance, concept of, 52
Statistical test, randomization test, 195,
        196, 200–201
Steeper power curves, 237
Stigler, Stephen, 210, 214–216
Stratified Block Randomization, 55, 56,
        57, 58
Streptomycin, 186
Streptomycin trial, in tuberculosis
        treatment (1948), 9–10
        Hill and, 3–5
Student's *t* test, 192, 200, 210–211
Suboptimal studies, consequences to
        using, 86–87, 86t
Subversion bias, 53
Symmetric summary receiver operating
        characteristic (SROC) curves,
        146, 151f–153f
Sympathetic bias, 29–37
        behaviors leading to, 31–32
        investigator preferences, 29–32
        motivations for, 29–30
        overview, 29
        preferences regarding predictability
                and sample size, 32–36,
                33f–37f
        fictional indifference curve, 35–36,
                36f
        fictional Nash equilibrium with
                unsympathetic investigator,
                36, 37f
        isoprofit curves, 32–33, 33f

iso-utility curves, 33–35, 34f
Nash equilibrium, 35, 35f, 36, 36f
response functions, 35, 35f
Synergistic solution, for selection bias,
        see Selection bias

T

Terminological clarification, 216–220
        overview, 216–217
        randomization tests
                computer-intensive tests and, 218
                exact tests and, 219
                exchangeability and, 219–220
                permutation tests *vs.*, 217–218
Test
        permutation, *see* Permutation tests
        randomization, *see* Randomization
                tests
        statistics, randomization, 195, 196,
                200–201
Third-order selection bias, 171, 172
        detection using Berger–Exner test,
                159–169
        background, 159–160
        bias testing, 163–164
        data analysis, 164
        discussion, 165–169
        limitations and strengths of
                method, 165–167
        methods, 160–164
        recommendation for further
                research, 169
        results, 164, 165t, 166t, 167t
        study scenarios and parameters,
                162–163, 163t
        trial simulation, 161–162
Treatment allocation predictability,
        76–79, 77f–78f
Treatment compliance, 31
Truthlikeness, 13, 13n
Tuberculosis
        streptomycin trial in treatment
                of (1948), 9–10
        Hill and, 3–5
Twain, Mark, 126
Twentieth-century (early) discussions,
        of bias and randomization in
        clinical research, 2–3

Two-arm BTR, 116, 117
Type I error, 53

U

Unbalanced designs, 191
Uncertainty(ies), 13
    ignoring, 13
"Unclear risk of bias," 84
Unequal allocation
    selection bias in studies with,
        89–120
        Blackwell-Hodges model for; *see
            also* Blackwell-Hodges model
        Blackwell–Hodges model for,
            91–97
        directional guessing strategy,
            97–107
        directional strategy with payoff
            threshold, 113–115, 115t–116t
        overview, 89–91
        PBR, 107–113, 108t–109t, 110f, 112f,
            112t
        in PBR in study with BTR, 19t,
            116–119, 117f, 118t
    trials, 90
Unit–treatment additivity model, 194
Unit–treatment interaction model,
    194–195
Unmasking (early *vs.* late), 133–135
Unrecognized dual threats, to internal
    validity, 123–137
    allocation concealment, dual threats,
        129–130
    betting odds *vs.* biasing allocation
        only with certainty, 130–131
    chronological bias *vs.* selection bias,
        128–129

false knowledge, propagation of,
    126–128
flawed trial evaluations translate into
    flawed trials, 124–125
malice *vs.* ignorance, 131–133
and Omnibus solutions, 135–136, 135t
overview, 123–124
unmasking (early *vs.* late), 133–135
Unrestricted randomization, 39–47;
        see also Randomization
    facts regarding, 45–47
    overview, 39–40
    selection mechanism
        MTI procedure induced by, 43–45
        MTI value for, 42–43
    theory and practice, 40–42
Urn Randomization, 53, 54, 55, 57, 58

V

Validity, by construction, 199–200
Verisimilitudes, 13, 13n

W

WBTR, *see* Wide Brick Tunnel
        Randomization (WBTR)
Welch, Bernard L., 213–214
Welch *t* test, 213
Wide Brick Tunnel Randomization
    (WBTR), 91
Wilcoxon-type statistics, 200
World War II, 3, 4
Wright, Almroth, Sir, 2, 5

Y

Yarnell, Harry E., 132